The Archeology of New Hampshire

The Archeology of New Hampshire

Exploring 10,000 Years in the Granite State

To Victor,

David R. Starbuck

Best wishes to a fellow student of the past!

David Starbuck

4-8-06

University of New Hampshire Press

Durham, New Hampshire

PUBLISHED BY UNIVERSITY PRESS OF NEW ENGLAND

HANOVER AND LONDON

UNIVERSITY OF NEW HAMPSHIRE PRESS

PUBLISHED BY UNIVERSITY PRESS OF NEW ENGLAND,

ONE COURT STREET, LEBANON, NH 03766

WWW.UPNE.COM

© 2006 BY UNIVERSITY OF NEW HAMPSHIRE PRESS

PRINTED IN THE UNITED STATES OF AMERICA

5 4 3 2 1

Unless otherwise indicated, illustrations are by the author.

Frontispiece: James and Donna-Belle Garvin examining pottery discovered at the Governor Wentworth Plantation site in Wolfeboro.

Library of Congress Cataloging-in-Publication Data

Starbuck, David R.
    The archeology of New Hampshire : exploring 10,000
    years in the Granite State / David R. Starbuck.
        p.   cm.
    Includes bibliographical references and index.
    ISBN-13: 978-1-58465-562-6 (paperback : alk. paper)
    ISBN-10: 1-58465-562-3 (pbk. : alk. paper)
        1. New Hampshire—Antiquities. 2. Indians of North
    America—New Hampshire—Antiquities. I. Title.
F36.S73   2006
974.2—dc22   2005032840

Contents

Appendixes

Preface

My goal is to present lively examples of what it's like to practice archeology in a small but dynamic New England state. I am presenting an overview of cultural changes throughout all of the major time periods in New Hampshire at many of the more important sites, and some of the more attractive and representative artifacts have been illustrated here. I am also including quite a few personal anecdotes, because, after all, archeology truly *is* an exciting field, a hobby and a profession that requires patience, hard work, optimism, and a passionate desire to discover all that is old.

Overviews do exist for the archeology that has been conducted in many other states, but this is the first such overview written for the state of New Hampshire. I know that this book will be enjoyed by lay audiences and students just as much as by professional scholars, and I have tried to keep theory and professional jargon to a minimum. Also, this book does not limit itself to just the prehistory of the state, as is so often done; it devotes just as much attention to the historical, industrial, and marine sites that have been explored in New Hampshire.

Perhaps I would not have attempted this broader perspective in a larger state that has seen more scholarship, but philosophically I believe this comprehensiveness reflects the thinking of many contemporary archeologists. Today most of us conduct research at prehistoric as well as historic sites, and we all believe that our students need exposure to the disparate research questions and techniques of both. Also, in an age when far more archeologists work in the private sector than in academia, those who conduct mitigation projects—popularly known as "contract archeology"—invariably work with the remains from *all* time periods. Only marine archeology is still practiced by a specialized few, given the inherent complexities and costs (and dangers!) of that field.

For those who aren't familiar with my work, I should explain that this book is a product of the 30 years that I have directed field projects and taught archeology courses in the State of New Hampshire. It is safe to say that I am still growing and learning as a researcher and a teacher, but hopefully this volume will share with readers a bit of the excitement that I've experienced while working in New Hampshire. The opportunity to conduct this research began when I arrived in southern New Hampshire in the fall of 1975, at which time I started directing excavations for Boston University at the site of the New England Glassworks in the town of Temple. Before that project ended in 1979, I had also begun a long-term survey and excavation project at Canterbury Shaker Village that continues intermittently even today.

Between 1981 and 1984, while first at the University of New Hampshire and later at Rensselaer Polytechnic Institute (RPI), I directed some of the very first SCRAP (State Conservation and Rescue Archaeology Program) excavations for the State of New Hampshire at two clusters of prehistoric sites on the Merrimack River, at Sewall's Falls and Garvin's Falls in Concord. I also occasionally directed some archeological contract projects during that period, including a very informative investigation in 1981 at some lithic workshops in the town of Belmont (NH31-20-5).

I co-directed excavations at the site of the Joseph Hazeltine pottery shop between 1982 and 1984. (Hazeltine was a redware potter who lived in the Millville district of Concord.) Next, between 1985 and 1988, while still at RPI, I co-directed excavations at the Governor Wentworth Estate in Wolfeboro, perhaps the most intact colonial plantation in New England. Most of these projects were conducted under the auspices of the New Hampshire Division of Historical Resources (DHR), typically with a local college as the co-sponsor.

In the 1990s, digs in New York State and Vermont increasingly filled my time, but I continued my Shaker research, and I was fortunate to excavate a small historic site in the backyard of Karl and Rika Schmidt's house in Orford, New Hampshire, in 1998–99. I am now (2005) working at the site of a 1730s frontier fort in Boscawen, the so-called "First Fort."

As my own work has progressed, I have watched many of the excavations directed by others in New Hampshire, and I have enjoyed talking with "old-timers" about the earlier digs they conducted in the state, prior to my arrival. I have tried to include a representative sample of these many projects, but not all, by any means. There simply are too many of these digs; also, many archeological sites in New Hampshire have not been sufficiently published on, or they have been described only in contract-funded reports, the so-called "gray literature" of archeology. I make no apologies for not including archeological sites in this book that have not yet been (and may never be) published on. Ironically, New Hampshire is a state where avocational archeologists have published as much (or more) than professional archeologists.

After 30 years of research, I owe a huge debt to colleagues, state officials, students, and other "diggers" who have worked with me and made it fun to be a scholar of early New Hampshire. I especially wish to acknowledge Linda Ray Wilson (DHR), David Hall (Harvard University), and James Wiseman (Boston University), the individuals who brought me to New Hampshire in the first place in 1975. I want to thank the many current and past officers and members of the New Hampshire Archeological Society (NHAS), who have been at the very center of archeological research over these many years. I also want to thank Mark Greenly and Jane Potter of the Collections Committee of the NHAS, who were very helpful in selecting artifacts to be illustrated in this book, and I am indebted to Patricia W. Hume, Chair of the NHAS Site Files Committee, who prepared appendix 2 based upon her extensive knowledge of the artifact collections housed throughout the state. Appendix 2 is an update of an earlier article first published by her and Donald Foster in 1994.

While preparing this book, I conducted several lengthy interviews and discussions with David Switzer, Don Foster, Dennis Howe, Davis Finch, Mark Greenly, Victoria Bunker, Jane Potter, and Mary Dupre, all of whom helped me to better understand the history and challenges of conducting archeology in New Hampshire. Their perspectives varied, but they all have a lifelong commitment to promoting a better understanding of New Hampshire's past. David Switzer, in particular, provided a great deal of help in preparing the chapter on marine archeology; without his many years of work as a maritime historian and archeologist, that chapter would not have been possible. Also, Victoria Bunker's careful reading of the manuscript has improved many parts of this book.

I especially wish to single out and thank Phyllis Deutsch, my editor at University Press of New England, whose wisdom and encouragement made this book possible. And I need to acknowledge my colleagues at Plymouth State University, whose support over the past 12 years gave me the necessary work environment in which to get my writing done. These include David Switzer, Kate Donahue, Grace Fraser, and Marcia Blaine, among others.

Other colleagues and friends who have helped over the years have included Richard Boisvert, the current New Hampshire State Archeologist, retired State Archaeologist Gary Hume, Dena Dincauze, the late Toni Howe, Eugene Winter, Mary Lou Curran, Karl Roenke, Betty Hall, Richard Borges, Roland Smith, Paula Dennis, Gini Miettunen, Ellen Savulis, David Lacy, Stephanie Tice, Merle Parsons, Maureen Kennedy, June Talley, Joseph Champlin, Sherry Mahady, Louise Luchini, Brownie Gengras, Jane Spragg, Linda Fuerderer, Judy Balyeat, Ann Doak, Robin Bagley, Millee Bolt, Brian Lombard, Ed McKenzie, Ken Rhodes, Paul Holmes, Dennis Chesley, Daniel Cassedy, Mary Cassedy, Ruth Cassedy, David Skinas, Stuart Wallace, Jim and Donna-Belle Garvin, Terry Kidder, Charles Harbage, Anne Giesecki, Robert Ewing, Billee Hoornbeek,

Howard Sargent, Charles Bolian, Scott Swank, Arletta Paul, Walter Ryan, and Cara Medyssey.

A very special feature of this book is the inclusion of a series of 31 superb line drawings prepared by Ellen Pawelczak of Darien, Connecticut, a most excellent illustrator. I am also indebted to Victoria Bunker, Richard Boisvert, Dennis Howe, Donald Foster, Mary Lou Curran, David Switzer, Karl Roenke, Daniel Cassedy, Gary Hume, Sheli O. Smith, Stephen Loring, Wesley Stinson, Davis W. Finch, Jean F. Topping, Katherine Donahue, and Gary Carbonneau for allowing me to use many of their photographs and line drawings in this book. It was Don Foster, Curator of the New Hampshire Archeological Society, who provided access to the collections and photographs of the NHAS that are included in this book. I would also like to acknowledge Ed O'Dell of Lake George, New York, who printed most of the photographs in this volume.

I would like to offer a very warm "thank you" to the family of Solon Colby, who allowed me access to the Colby Collection, which is on loan to the NHAS and is on exhibit in the Anthropology Museum at Phillips Exeter Academy. The kindness of Carolyn Colby Brouillette and Martha Brouillette-Martel has made it possible for me to include examples of some of the very best Colby artifacts in this volume.

Finally, I want to thank every last student and volunteer who has dug with me in New Hampshire over the past 30 years. It was really *your* work that made this book possible!

July 2005   D.S.

Note on the Text

Archeology has long had two spellings, reflecting tradition, personal preference, and the decision of the U.S. Government Printing Office to modernize the spelling of the word "archaeology" by dropping the second "a." It is amusing to some and puzzling to others that archeologists can be positively passionate about what is the "right" way to spell the name of our field! I have chosen to use the "archeology" spelling in this book, but in cases where archeology appears as part of a formal title, such as "The New Hampshire Archeological Society" or the "New Hampshire State Archaeologist," I have preserved the original spelling of the word.

The Archeology of New Hampshire

Introduction

The History of Archeological Research in New Hampshire

Background

T HIS BOOK IS about archeology as practiced in New Hampshire. I believe that it is long overdue, and I hope to demonstrate that discoveries made in New Hampshire can be just as interesting as anything that explorers might find inside tombs, temples, or pyramids in distant lands. I will present overviews of the changing lifestyles enjoyed by the people of New Hampshire over the past ten thousand years or more, and I will describe many—but by no means all—of the archeological sites that have been meticulously dug, measured, and sifted by avocational and professional archeologists over the past century (fig. Int.1). This will hopefully be more than a mere collection of "facts" about the past. Rather, I will also be telling stories and anecdotes about the archeologists, what we hope to learn, and the techniques we have used to glean information from the past. After all, if we can come to understand why grown adults would give up safe and lucrative "real" jobs for the chance to dig in the dirt, then maybe, just maybe, we will learn what makes the past meaningful and exciting to so many of us.

But I want to begin with a cautionary statement. One of my first and strongest memories about archeology in New Hampshire comes from the 1970s, shortly after I began doing archeological research in the state. Archeologists at the University of New Hampshire in Durham conducted a survey of schoolteachers in New Hampshire and learned that over 70 teachers were busily digging archeological sites (chiefly cellar holes) with their students every year. None of these sites was ever published upon, and it was clear that sites were being destroyed at an alarming rate, even by those who were genuinely well-intentioned. It cannot be overemphasized that no archeological site is expendable, and nothing should be dug by those who lack the proper training or want students to "get their hands dirty." We archeologists dig in order to answer questions about the past, to understand how past peoples made their

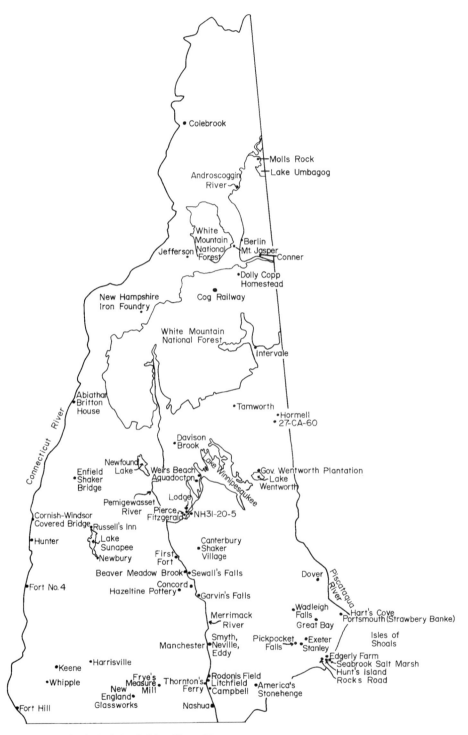

FIG. INT.1. Archeological sites in New Hampshire.

decisions about where to live or how they obtained food and other resources, and to determine why and how cultures changed over time.

Having said that, what *is* the proper role of archeology in our society, and how does one get adequate training so that the remains of the past will be able to educate and fascinate everyone who wants to learn more? There really is no substitute for a formal academic training, and university courses in archeology and anthropology have helped thousands of Americans develop a greater appreciation for the past and for the wonderful diversity of human cultures that has shaped the complex nature of our society today. Still, most students who have been trained in archeology do not go on to pursue careers in this discipline. There simply are not enough jobs for everyone who has an interest, but many have been able to satisfy their love for archeology by working seasonally on archeological digs, by attending meetings of organizations such as the New Hampshire Archeological Society (NHAS) or the New England Antiquities Research Association (NEARA), or by subscribing to the many "popular" magazines that cover our field, such as *Archaeology, Archaeology Odyssey, American Archaeology, National Geographic, The Biblical Archaeology Review,* and *Dig* (for kids ages nine and up).

Archeological digs and laboratories are always looking for serious volunteers. This is, after all, a very labor-intensive field, and a great many professionals trained in other fields have been able to make meaningful contributions to archeology with their time and ideas. The role of the serious avocational must never be underestimated, and avocational archeologists in New Hampshire have been some of the most dedicated workers in our field (Boisvert 1994b).

The Development of New Hampshire Archeology

There are several candidates for "the first archeologist" in New Hampshire, but early figures such as Wentworth Cheswill, Sebastian Griffin, and Samuel Sewall Parker deserve special mention (Chesley 1978–1979, 1982; Chesley and McAllister 1981), as do Jeremy Belknap (1792), Chandler Potter (1856), and Frederick Putnam (1873). All of these pioneers wrote about Native Americans and the chance discoveries of "Indian relics" in New Hampshire. Out of this group of early scholars, Dennis Chesley and Mary Beth McAllister have given Wentworth Cheswill the title of "New Hampshire's first archeologist" because of an archeological report that Cheswill wrote in about 1790 (Chesley and McAllister 1981).

Nevertheless, it was perhaps Warren Moorehead (1930, 1931) of the Robert S. Peabody Foundation in Andover, Massachusetts, who deserves to be recognized as the first "real" archeologist to work in New Hampshire, given the systematic nature of his surveys along the Merrimack River and his excavations at the Smyth site in Manchester in 1930. Soon afterward, an avocational archeologist, Harlan Marshall, published nationally (1942) on three Indian sites in the Manchester area, and in 1947 Elmer Harp, Jr., of Dartmouth College initiated a prehistoric survey in the Hanover area as part of the coursework at that college. To Harp no doubt falls the distinction of being the first professionally trained and college-employed archeologist to do field work in New Hampshire.

It was in November of 1947 that the first members of what became the NHAS met in Exeter (Holmes 1972). Soon afterward, the group met at the Manchester Historic Association, as was reported in the *Morning Union* on December 3, 1947:

> The New Hampshire Archeological Society was organized last Saturday afternoon, when a group of people from different parts of the state, interested in the study of Indian lore, gathered at the Manchester Historic association building on Amherst street.
>
> Laurence M. Crosbie of Exeter, who called the meeting to order, explained that the new organization had been suggested in order that those in the state, who are making a study of Indian life and remains, may have some meeting place, where information can be explained and projects to collect and preserve the evidences of early civilization can be formed. He pointed out that similar societies existed in all but three of the Atlantic Seaboard states.
>
> A constitution was adopted and officers were elected. Membership, as reported at the first meeting, numbered 45. Mr. Crosbie was elected president; Frank O. Spinney, director of the Manchester Historic association, James T. Schoolcraft, University of New Hampshire, and Henry Phillips, Jr., Exeter, were chosen as vice presidents; Elmer Harp, Jr., assistant curator, Dartmouth College Museum, Hanover, was named secretary-treasurer.

Since its inception in 1947, the NHAS has been the primary institution in the state devoted to archeological research, publication, and public education. Through the publication of a newsletter and a journal, *The New Hampshire Archeologist*, the society has sought to promote a better understanding of the prehistory and history of the state of New Hampshire.

Under the early leadership of Laurence Crosbie (fig. Int.2), the society held its first dig at the so-called "Ossipee Indian Mound," which is a glacial deposit in the town of Ossipee (Brown 1950). This was followed by several years of work at Clark's Island in Lochmere under the direction of Howard Sargent. Sargent had recently arrived in New

FIG. INT.2. Laurence M. Crosbie, the founder and first president of the New Hampshire Archeological Society, in 1947. Courtesy of the Anthropology Museum, Phillips Exeter Academy.

★ Howard Sargent (1922–1993), an Early Professional

For many years, Howard Sargent was the face of New Hampshire archeology, one of the few dedicated "professionals" in the state who was well known to archeologists throughout the Northeast. Trained at Yale University (B.A. 1950) and the University of Michigan (M.A. 1951), Howard conducted significant projects at the Hunter site, Sumner Falls and other locations on the Connecticut River, the Smyth site in Manchester, Weirs-Aquadocton in the Lakes Region, Garvin's Falls on the Merrimack River, Lake Umbagog, the Tenney site in Hillsboro, Cartagena Island in Bedford, the Russell's Inn site in Sunapee, and a host of other sites in practically every corner of the state.

I remember fondly how in 1975, when I was about to begin my own work in New Hampshire, Irving Rouse at Yale University proudly told me that another one of his students—Howard—had been digging in New Hampshire for years. I was to be the second of Rouse's students to work in New Hampshire!

Howard mentored many of New Hampshire's archeologists and was invariably kindhearted, patient, and gracious. While some of his work was done through the auspices of the New Hampshire Archeological Society, which awarded him the Chester B. Price Award in 1971 for outstanding service, much was also done through the several colleges where he taught, most notably Franklin Pierce College in Rindge (1967–1973 and 1978–1988) and Nathaniel Hawthorne College (1963–1967). Howard also worked with various cultural resource management firms. While perhaps best known within New Hampshire for his many archeological surveys and excavations, Howard was especially known in academic circles for the professional journal he founded in 1971, *Man in the Northeast* (later renamed *Northeast Anthropology*), which he edited until 1982.

More than any other professional archeologist in the history of New Hampshire, Howard was an expert in public education. He lectured unceasingly to all manner of public groups over the many years he worked in New Hampshire, and he included vast numbers of volunteers and students in his projects. Howard was also named Teacher of the Year by the Franklin Pierce College Student Senate in 1988 (Snow 1993). When Howard passed away in 1993 at the age of 71, he left the state much better known than when he arrived, the result of many years of excited, passionate research.

Today the Sargent Museum of Archaeology, located in Manchester, continues Howard's legacy. See <www.sargentmuseum.org>. For many years under the direction of Wesley Stinson, the Sargent Museum houses most of Howard's artifact collection and features talks and public programs on New Hampshire archeology. So now, through others, Howard continues to have a very positive impact upon the people of New Hampshire.

Howard Sargent at the Pickpocket Falls site in Exeter. Courtesy of the New Hampshire Archeological Society.

FIG. INT.3. Dr. Eugene Finch (right) and Howard Sargent at the Pickpocket Falls site in Exeter in 1957 or 1958. Eugene Finch was a very serious avocational, and together with his son, Davis, he participated on many digs in the 1950s and 1960s. The Edgerly Farm site in Hampton Falls was probably his favorite site, and Davis has many happy memories of digging with his father (Davis Finch, personal communication, January 19, 2005). Eugene's high professional standards and many publications made him one of the best archeologists of his day. Courtesy of the New Hampshire Archeological Society and Davis Finch.

Hampshire, having received his B.A. from Yale University and his M.A. from the University of Michigan. Sargent's impact upon archeology in New Hampshire was immediate and profound. In the years that followed, the field work conducted by NHAS members often consisted of aiding Sargent in his many digs throughout the state (see the box "Howard Sargent (1922–1993), an Early Professional").

However, a large and very capable group of avocational archeologists also developed at that time, and prominent among them were Solon Colby (see the box "Solon B. Colby (1900–1971), an Early Explorer") and his brothers Perley and Robert, Eugene and Davis Finch (fig. Int.3), Chester Price, Frederic Burtt, Clyde F. Berry, Percy S. Brown, William Fisher, George Prindle, Paul Holmes, William White (fig. Int.4) and, of course, Laurence Crosbie. Part of Crosbie's artifact collection is on display at Phillips Exeter Academy, and it was a young Howard Sargent who used the artifacts to create exhibits in about 1953 (Donald Foster, personal communication, January 19, 2005). These pioneers were joined by Eugene Winter (fig. Int.5) in 1965, at which time Winter directed excavations for the NHAS at the Garvin's Falls site in Concord and then at the Smyth site in Manchester. Other digs were held at the Great

★ Solon B. Colby (1900–1971), an Early Explorer

The most distinguished family in the history of New Hampshire archeology is, in the eyes of many, the Colby family of Bow, New Hampshire, and first among the Colbys was Solon B. Colby, popularly known as "Mr. Indian of New Hampshire." Solon and his brothers, Robert and Perley, assembled the most impressive collections of prehistoric artifacts ever found in New Hampshire, chiefly from the Concord, Bow, and Suncook area on the Merrimack River. Many of the artifacts in the Colby Collection were recovered from the Smyth site, Garvin's Falls, Sewall's Falls, and the point where the Suncook River flows into the Merrimack River. Nothing in the Colby

Collection was bought from any other digger—everything was found either by Solon or one of his brothers (Robert Colby interview, December 16 and 19, 1981).

Solon served as president of the New Hampshire Archeological Society from 1964 to 1968. His love for the Indians of New Hampshire was almost legendary, and his

Examples of ground stone tools in the Colby Collection (clockwise from upper left): a fully grooved ax, a gouge (may have come from Bow), a plummet, and an atlatl weight.

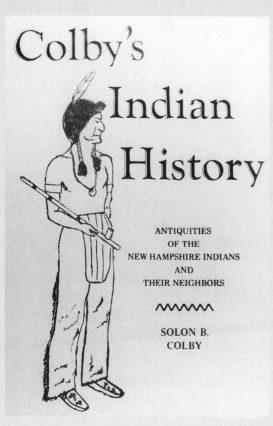

Cover of *Colby's Indian History*.

ANTIQUITIES
OF THE
NEW HAMPSHIRE INDIANS
AND
THEIR NEIGHBORS

〰〰〰〰〰

SOLON B.
COLBY

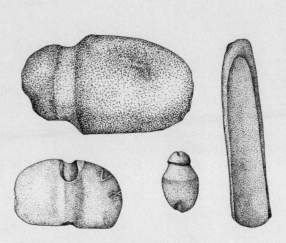

The same artifacts illustrated by Ellen Pawelczak.

many accounts of Indian events in New Hampshire were published posthumously as *Colby's Indian History* (1975). His efforts to share his knowledge with the public, and his publication of what he had learned, were laudatory, and the sites and artifacts he found are now of considerable benefit to science. In an obituary prepared for *The New Hampshire Archeologist*, his friend J. Frederic Burtt noted that "'Mr. Indian of New Hampshire' brought breath and life to them as he described the native American and the aboriginal origins of the people who once inhabited the Grants of New Hampshire." He continued, "Recognizing the great archeological potentialities of the Smyth Site at Manchester, and Garvin's Falls at Pembroke, he worked for and received the necessary permissions for the opening of these important Indian settlements. Each has proven to be of great scientific worth in the enlargement of archeological knowledge in New England" (Burtt 1971).

In 1981 the family of Solon Colby placed his collection of Indian artifacts on loan to the New Hampshire Archeological Society "for curation and educational display," and many of the finest artifacts are now on display in the Anthropology Museum at Phillips Exeter Academy. Some of the most impressive of these artifacts are depicted in this book, reflecting the outstanding quality of the Colby Collection. (See appendix 4 for some of the projectile point types represented in the Colby Collection.) If he were still with us, hopefully Solon would be pleased at the continuing interest in his work.

FIG. INT.4. William J. White, pictured with part of his collection. Bill White is a self-taught archeologist, now in his eighties, who assembled a huge collection that is now at Phillips Exeter Academy. It was he who introduced Don Foster to the Stanley site. Photograph by Karen Singer. Courtesy of the Anthropology Museum, Phillips Exeter Academy.

FIG. INT.5. Eugene Winter helping with the dig at Weirs Beach in 1976. Gene's prowess as an outstanding avocational archeologist and educator earned him the 2005 Crabtree Award from the Society for American Archaeology.

Bay site in Greenland, the Indian Fort in Lochmere, Pickpocket Falls in Exeter, Litchfield Farm in Litchfield, the Edgerly Farm in Hampton Falls, and elsewhere. Intensive field work throughout the state resulted in the formation of chapters of the NHAS in the 1950s, the Monadnock Chapter, the Sea Coast Chapter, and the Dartmouth Sunapee Chapter, but these were short lived.

The antiquarian work of the nineteenth century had been replaced by regional surveys in the early twentieth century, and then the formation of the NHAS in the mid-twentieth century led to substantive excavations and publications. Still, in 1950 the NHAS had only 48 members. At that time, most categories of dues cost just $1 (institutional, educational, active senior, active junior), with a sustaining membership going for $5 and a life membership for $50. Members were expected to take their own equipment to society digs, and during the early years everyone was allowed to keep the artifacts that he or she found. Later, starting in 1963, the Chester B. Price award was established "for outstanding contributions to the cause of archeology in New Hampshire." All of the recipients of this award are listed below:

1963	J. Frederic Burtt	1971	Howard Sargent
	Solon Colby	1980	Kenneth Rhodes
	Eugene Finch	1981	Eugene Winter
	Peter McLane	1982	William White
	George Prindle	1985	Dennis Howe
1970	Paul E. Holmes		Antonett Howe

1986	Edward McKenzie	1994	Patricia Hume
1987	David R. Starbuck	1997	Jane Potter
1989	Louise Tallman	1999	Justine "Brownie" Gengras
1992	Donald Foster	2002	Gary W. Hume
1993	Charles Bolian	2005	David Switzer

Interestingly enough, no woman received the Chester B. Price Award until 1985, and that was Antonett "Toni" Howe. Almost all of the original NHAS diggers had been men, and nationwide it was not until the 1970s that large numbers of women began training to become archeologists. This modest beginning is in very sharp contrast to today; women now occupy many of the most prominent positions in the field of archeology.

A pattern of small, almost family-style excavations lasted up until 1967, 1968, and 1969, at which time larger excavations were conducted at the Smyth and Neville sites in Manchester, where Warren Moorehead had previously dug and found burials in 1930 (fig. Int.6; Moorehead 1931:62). These were prominent, well-stratified fishing stations on the east bank of the Merrimack River at Amoskeag Falls, which are now surrounded by the modern city of Manchester. These sites were arguably the first in New Hampshire where archeologists were able to view rich, deeply buried deposits capable of giving New Hampshire's prehistory a prominence throughout the Northeast. Field work at the Amoskeag sites by Peter McLane, Ken Rhodes, Emmanuel Valavane, Eugene

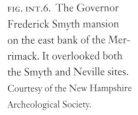

FIG. INT.6. The Governor Frederick Smyth mansion on the east bank of the Merrimack. It overlooked both the Smyth and Neville sites. Courtesy of the New Hampshire Archeological Society.

FIG. INT.7. Dr. Charles Bolian (right center) conferring with Anne Giesecke at Weirs Beach in the summer of 1977.

Winter, Howard Sargent, and others marked a huge turning point in New Hampshire archeology. Subsequent publication of the long chronological sequences at Neville (Dincauze 1971, 1975, 1976) and Smyth (Foster, Kenyon, and Nicholas 1981; Kenyon and Foster 1980; Kenyon 1985b) helped to give New Hampshire archeology the attention and respect it deserved among archeologists outside the state.

While Howard Sargent and Elmer Harp were the only college-based archeologists working in New Hampshire prior to 1970, this began to change after Charles Bolian was hired by the University of New Hampshire (UNH) in 1972. Bolian's subsequent work, digging from 1974 to 1975 at the Rocks Road site in Seabrook (Robinson and Bolian 1987) and at Weirs Beach from 1976 to 1979 (fig. Int.7; Bolian 1976–1977, 1980), very quickly made UNH a leader in New Hampshire archeology. Within a few years, most of the digging conducted by avocationals was in concert with Bolian, Sargent, or one of the other professional archeologists who were just beginning to arrive in the state.

At the Rocks Road site, Bolian was assisted by Brian Robinson, Ed McKenzie, Eugene Winter, Evelyn Fowler, Marjorie Chandler, Paul Holmes, Eugene Finch, Hans Barlow, Billee Hoornbeek, Mary Dupre, many other members of the NHAS, other students from UNH, and Donald Foster and classes from Phillips Exeter Academy. Quite a few of the Seabrook workers continue to be active in the field today. During this period Phillips Exeter Academy developed even closer ties with the NHAS. Don Foster, who had been hired in 1973 to teach cultural anthropology at the academy, led students in excavations at the Stanley site on the Exeter River (fig. Int.8; Foster 1982). Foster assumed the roles of NHAS editor, president, and, later, curator. The NHAS has been extremely fortunate that in recent years its archeological collections have been curated at the academy under Foster's supervision (figs. Int.9 and Int.10).

After 1970, newly passed environmental legislation was also starting to have an impact on how archeology was funded, and the requirements of cultural resource management surveys in the 1970s (conducted

FIG. INT.8. Dr. Donald Foster (right) and a Phillips Exeter student digging at the Stanley site in the 1980s. Courtesy of the New Hampshire Archeological Society.

FIG. INT.9. Dr. Donald Foster (left) meeting with Mark Greenly and Jane Potter of the Collections Committee of the New Hampshire Archeological Society.

as part of the environmental review process) necessitated that the state
finally hire an archeologist to oversee the protection and recording
of endangered sites. Thanks to the urging of Sargent, Bolian, and the
NHAS, a part-time position was created in 1976 in the State Historic
Preservation Office (SHPO) in Concord (now better known as the New
Hampshire Division of Historical Resources). Gary Hume was hired
as the first SHPO archeologist, and within just a few years the position
became full-time. Still, it took several additional years before the State
of New Hampshire was willing to add the title of "State Archaeologist"
(see the box "Dr. Gary Hume, New Hampshire's First State Archaeologist [1976–2002]").

Several years later, in 1982, Hume created the State Cooperative
Regional Archaeology Plan (SCRAP), later retitled the State Conservation and Rescue Archaeology Program. Since then, this state-sponsored
program has provided field and laboratory opportunities, under professional supervision, for avocationals throughout the state of New Hampshire (see the box "Antonett Howe (1937–1992), an Avocational Archeologist for Our Time"). Several of the most senior members and officers
of the NHAS received their first training on SCRAP projects at Sewall's
Falls and Garvin's Falls in the early 1980s (Starbuck 1982a, 1985a), and
SCRAP has continued to provide public participation programs and
an annual summer field school up to the present day (fig. Int.11). One
outcome is that current NHAS members rarely conduct projects on

FIG. INT.10. New Hampshire
artifacts exhibited in the
Anthropology Museum at
Phillips Exeter Academy.

★ Dr. Gary Hume, New Hampshire's First State Archaeologist (from 1976 to 2002)

Dr. Gary Hume.
Courtesy of Gary Hume.

In July of 1976, Gary Hume was hired into a joint staff position between the New Hampshire State Historic Preservation Office (SHPO) and the University of New Hampshire. By 1979 this had evolved into a full-time position at the SHPO, and Gary continued in this role until his retirement in 2002. The arrival of a state archaeologist could, perhaps, be compared to "the taming of the West" in the sense that it meant trying to satisfy diverse constituencies who were already used to doing things a certain way. A state archeologist needs an excellent knowledge of state and federal regulations but must also work well with the professional and avocational archeological communities, with government officials, with the Native American community, and with developers and the business community, all the while trying to maintain high archeological standards. To be the very first state archeologist in New Hampshire, charged with educating all of these groups, could not have been an easy task.

Before Gary became New Hampshire State Archaeologist, he had been a college professor at the American University and at George Washington University, and he had also been directing excavations in Iran with his wife, Valerie. Later, Gary worked as an archeological contractor. Beginning in 1976, his primary responsibility was to enforce state and federal laws that protect archeological sites, in effect "leading the charge" when it came to ensuring that developers and others would have to take archeological sites into consideration and not simply remove them. It must sometimes have been frustrating to no longer conduct his own personal field work so that he could monitor the work of others, in effect managing vast amounts of paperwork instead of digging and analyzing artifacts! Still,

every state archeologist has the opportunity to make lasting contributions, and Gary's innovative approach in New Hampshire was to create the State Conservation and Rescue Archaeology Program: "In 1981, New Hampshire became the first state to authorize through legislation the training, certification and use of avocational archeologists" (Hume 2003–2004: 7). The training and certification of large numbers of avocationals led to a significant broadening of public participation in New Hampshire archeology, and this no doubt was Gary Hume's greatest legacy to the people of New Hampshire. There were other successes, too, including overseeing the return of many sets of human remains to the Abenaki Nation of Missisquoi (under NAGPRA, the Native American Graves Protection and Repatriation Act); conducting historical research for the Monson Rural Archaeological District in Milford-Hollis; and anthropological research for the Abenaki Indian Shop and Camp in Intervale (Hume 1991).

Most of the tasks of a state archeologist are anything but glamorous, and some of Gary's other activities included occasionally battling with developers, helping to draft legislation to protect sites, delivering public lectures, attending innumerable public and private meetings, and constantly visiting archeological sites all over the state of New Hampshire. For students contemplating a career in archeology, it may appear that Gary's tasks lacked the "fun" found in digging, analysis, and publishing, but the essential activities performed by the state archeologist make it possible for all of the rest of us to do our work.

★ Antonett Howe (1937–1992), an Avocational Archeologist for Our Time

The real backbone of New Hampshire archeology is its volunteer component, the avocational archeologists who devote countless hours to digging, laboratory work, collections analysis, and writing. Within the volunteer community, one of the most generous and devoted archeologists was Antonett "Toni" Howe, who died from renal cell carcinoma in September of 1992. She was just 55.

Toni had first been introduced to archeology in 1981, when she participated in a summer field school under my direction at Sewall's Falls in Concord. In the years that followed, she helped to train a great many other volunteers and students of archeology, and she enjoyed what she was doing so much that she herself went back to school to receive a B.A. in anthropology from the University of New Hampshire in 1986. Together with her husband, Dennis, she received the Chester B. Price Award from the New Hampshire Archeological Society in 1986, and she was named "Avocational Archaeologist of the Year" by the New Hampshire Division of Historical Resources in 1987.

Toni taught other prospective archeologists how to dig, she helped reassemble and analyze artifacts in the SCRAP Laboratory in Concord, and she took part as a digger and field supervisor on several projects in New Hampshire and New York. She also excelled as an avocational in that she presented talks at archeology meetings, she attended professional archeological conferences, and she published an article about her work that appeared in *The New Hampshire Archeologist* (1989). This last step was a crucial one, for as she learned to publish, she truly became a "professional" in her work. One of the projects that I will forever associate with Toni was our excavation of the Governor Wentworth Plantation in Wolfeboro. She dug, she processed artifacts, and she suffered greatly from the heat inside the mansion cellar hole. (It was a real heat trap!) But she persevered.

In a great many ways, Toni's work and life symbolized what we all should be doing as archeologists. She loved people, she loved archeology, and she shared her knowledge.

Antonett Howe looking up from her pit at Garvin's Falls in the summer of 1982.

Archeology requires patience and teamwork—even in the laboratory! Here SCRAP volunteers are reassembling a giant case bottle discovered at the Governor Wentworth Plantation in Wolfeboro. Toni is on the far left.

FIG. INT.11. SCRAP field crew, 1998 Field School, Jefferson III site (27-CO-30), Jefferson, New Hampshire. Courtesy of Richard Boisvert, NHDHR.

FIG. INT.12. Dr. Richard Boisvert, 2004, at the Jefferson II site (27-CO-29), Jefferson, New Hampshire, with Lucy Burris, SCRAP Avocational Volunteer of the Year for 2004. Courtesy of Laura E. Jefferson, NHDHR.

FIG. INT.13. The White Mountain National Forest works in partnership with organizations such as the Girl Scouts of the United States of America to educate the public and accomplish required archeological excavation in advance of forest projects. Forest Archaeologist Karl Roenke (right) and Girl Scouts conduct test excavations at the Dolly Copp Homestead site (in use from about 1830 to 1880) near Gorham, New Hampshire. Courtesy of the White Mountain National Forest.

their own; instead, most of them take part on SCRAP projects, typically under the leadership of the present New Hampshire State Archaeologist, Richard Boisvert (fig. Int.12). Some of these avocationals have also been employed part- or full-time on cultural resource management projects.

Another venue for research in New Hampshire has been provided by the White Mountain National Forest (WMNF), which began its Cultural Resource Program in 1979 with the hiring of its first Forest Archaeologist, Billee Hoornbeek. She was followed in 1988 by Karl Roenke, who continues today as Heritage Resource Program Leader for the WMNF (fig. Int.13). Standardized project reports are produced by professional archeologists and Heritage Resource Paraprofessionals in accordance with legal requirements. Heritage Resource Paraprofessionals are field-going United States Forest Service employees who have undertaken a formal training program in Heritage Resource Management. Their goal is to understand the layers of land-use history and to use that data to provide logical and defensible recommendations for the management of our cultural heritage.

Management programs are clearly one of the latest ways to better understand and protect cultural resources in New Hampshire, and undeniably most excavations today are the result of state and federal laws. Still,

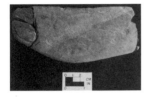

FIG. INT.14. A ground stone ulu (site 27-CA-112) found during a highway bypass study in North Conway in the mid-1990s. This is an example of the types of artifacts that can be found while conducting cultural resource assessments in New Hampshire. Courtesy of Victoria Bunker.

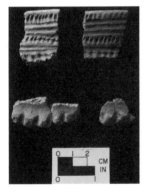

FIG. INT.15. Sherds from a collared and decorated prehistoric vessel (site 27-CA-102) found in North Conway. Courtesy of Victoria Bunker.

modern archeology continues to include the occasional "pure" research project, conducted principally to answer academic questions about specific cultures. Also, these efforts are accompanied by SCRAP-sponsored surveys and excavations, and by college field schools offered by Franklin Pierce College and Plymouth State University. All of these efforts do, of course, have research components, but the need to avoid or document threatened sites is now the driving force behind most New Hampshire archeology. This definitely does not make archeology any less exciting, but it points out the modern awareness that we need to protect sites that would otherwise be lost to developers, highways, and treasure-hunters. "Progress" has oftentimes not been kind to archeological sites, but the environmental review process has led to a host of recent excavations that have added to our knowledge of early New Hampshire. For example, the many projects conducted by the firms Victoria Bunker, Inc., and Independent Archaeological Consulting have resulted in cultural assessments in many parts of the state and in the preparation of excellent publications afterward (figs. Int.14 and Int.15).

There has also been a marked increase in the number of radiocarbon dates obtained from prehistoric sites in New Hampshire. A comprehensive listing of all dates obtained through 1995 was compiled in an article by Justine Gengras (1996), who subsequently led a radiocarbon dating project through the auspices of the NHAS. This project was supported with Intermodal Surface Transportation Enhancement Act funds from the U.S. Department of Transportation, supplemented with contributions from the NHAS and the NEARA. Consequently, many more dates are now available to scholars, helping to place New Hampshire prehistory on a much sounder chronological footing.

As we look to the future, archeology will no doubt continue to be an important field for enhancing our knowledge of New Hampshire's past. The public will always have opportunities to assist in research, buy books, attend lectures, and protect the prehistoric and historic sites that are located in many of the "backyards" of New Hampshire. Archeological sites are not a renewable resource, and it is up to *our* generation to learn as much as we can before the past is lost forever. Every archeologist has the potential to become a "storyteller," sharing what we have learned with all who will listen.

Part I

The First People of New Hampshire

★ When the first European explorers and fishermen arrived on the coast of New Hampshire, they encountered a rich Native culture that had been evolving for more than ten thousand years. New Hampshire in the late 1500s was occupied by peoples speaking the Western Abenaki language, and these included the Penacooks and Winnipesaukees along the Merrimack River and the Sokoki along the upper Connecticut River. What the Europeans could not know was the great time depth of these First Americans, nor were they aware of the stages of cultural development experienced by the original people of New Hampshire. But it is undeniable that Native peoples of every time period had adapted quite successfully to their environment(s) and made good use of available resources, locating themselves where diverse natural resources were easily available or where different ecological zones—"microenvironments"—came together.

It has taken many years of systematic excavations by prehistorians, coupled with hundreds of carbon-14 dates, to establish the sequence of prehistoric cultures in the eastern United States that is generally accepted among archeologists today. New Hampshire was a part of much broader cultural trends that characterized the entire culture area east of the Mississippi, where everyone adapted to similar lake, forest, and riverine environments and had similar sources of subsistence. Pioneering efforts by William Ritchie in New York State and James Griffin in Michigan, coupled with less-well-known research by archeologists working in other states, has resulted in the outline of time periods presented in table I.1. These divisions reflect somewhat arbitrary boundaries defined by artifact assemblages, and they represent changing environments and socioeconomic conditions. As archeological investigations have been conducted, these stages of development have

TABLE I.I. Major Stages of Cultural Development

Period	Time span	Subsistence patterns/development
Paleo-Indian	10,000–8000 B.C.	Highly nomadic, hunting, gathering, post-glacial adaptations
Archaic	8000–1300 B.C.	Hunting, fishing, gathering
Transitional	1300–1000 B.C.	Use of stone bowls, hunting, fishing, gathering
Woodland	1000 B.C.–A.D. 1600	Ceramics, horticulture, village life
Contact	A.D. 1600–Present	Horticulture, village life, trade with colonists

been refined to reflect localized settlement patterns, stylistically similar artifact types and attributes, and culturally distinctive characteristics within particular areas.

While there are now a growing number of archeological sites in North and South America that appear to pre-date 10,000 B.C., we nevertheless cannot place human populations in New Hampshire before about 10,000 or 9500 B.C. for the simple reason that glacial conditions would not have permitted human entry into the state. Consequently, we begin the prehistoric occupation of New Hampshire with the well-documented Paleo-Indian period when people were living on lake shores and river terraces, and on the edges of pro-glacial lakes, and continue through the Archaic and the Woodland periods to the point of European contact at about A.D. 1600.

There are several excellent syntheses that document these cultural developments on a national scale, with Brian Fagan's (2005) being one of the best. Regionally, the prehistory of the northeastern United States is well-covered by Levine, Sassaman, and Nassaney (1999), Trigger (1978), Snow (1980), Funk and Hayes III (1977), and Spiess (1978). New Hampshire has received very little coverage in each of these national and regional overviews, perhaps because most of the research conducted in New Hampshire has been published in relatively "local" publications that are seemingly little consulted by outside authors. However, there is one short, excellent overview of New Hampshire's prehistory, written by Victoria Bunker (1994). Bunker is especially good at reviewing the factors that determined human site location, noting that the first residents of New Hampshire "did not roam aimlessly across the landscape but made choices on when and where to settle and on the types of activities best suited to each place" (1994: 20).

The publications of the NHAS are the primary medium through

which the state's prehistory has been reported, but occasional articles have also appeared in the journals *Archaeology of Eastern North America*, *Man in the Northeast* (now *Northeast Anthropology*), and elsewhere. Also, there are some excellent Ph.D. dissertations and master's theses that have dealt with aspects of New Hampshire prehistory, including Kenyon (1983), Curran (1987), and Cassedy (1984).

This book follows conventional practice in devoting separate chapters to the Paleo-Indian period, the Archaic period, and the Woodland period, but this is oftentimes a difficult approach, because most of New Hampshire's prehistoric sites are multicomponent. In other words, they experienced reoccupation at various times throughout the Archaic and Woodland periods, and it is awkward to trace different occupations of a single site through successive chapters. This is the inevitable result of peoples during different time periods generally seeking the same natural resources, in the forms of access to water for transportation, fishing, and associated flora and fauna; well-drained soil; good lithic material and clay; and southern exposures for optimal warmth and sunlight.

The approach followed in each chapter is to begin with a national perspective, follow this with a regional overview, and then focus upon selected New Hampshire sites and patterns. This inevitably means that sites that have been published upon more intensively will receive a more thorough treatment, and some sites will not or cannot be covered at all for lack of available information. I am fortunate to be able to present three of these sites from a personal perspective (Sewall's Falls, Garvin's Falls, and NH31-20-5 in Belmont) because I conducted fairly extensive projects at them and so can provide more meaningful insights and interpretations. I regret that there are a host of other New Hampshire prehistoric sites that I have not been able to include here, but there are simply too many to present them all adequately.

Chapter 1

The Paleo-Indian Period

The Big-Game Hunters

The First People to Enter the Americas

THE MOST exciting topic in American archeology today is easily the question of origins: When did the first people enter the New World, by what route(s), what types of tools did they bring with them, and why have we found so little incontrovertible evidence for them? For many years the traditional view was that around 10,000 or 9500 B.C., small groups of hunters began to cross the Bering Straits into what is now Alaska and the Yukon, following and hunting Pleistocene megafauna, which in turn were feeding on vegetation that had recently taken root in an ice-free corridor between adjacent sheets of melting ice. Few early sites were ever discovered along this route, but that was not seen as a problem, because retreating glaciers had no doubt scoured the land surface, removing any traces of human activity.

This paradigm, now seen as the most cautious, had its beginnings in 1927, when the American Museum of Natural History sponsored the excavations in Wild Horse Arroyo at Folsom, New Mexico. Earlier excavators had discovered projectile points believed to be extremely old, but in 1927 a fluted projectile point was found between the ribs of a species of bison (*Bison antiquus*) known to have been extinct since the end of the last Ice Age. The Folsom find made it possible for most American scholars to gradually come to accept the great antiquity of the First Americans. Soon after, discoveries made in 1932 at the site of Blackwater Draw near Clovis, New Mexico, led to the naming of the distinctive "Clovis point," the fluted spear point that has long been seen as the earliest clear indicator of human presence everywhere it has been found in the New World. While Folsom-style points have proven to be relatively local within the Southwest, Clovis points have been found throughout North and South America. Because the Pleistocene megafauna died off within a very few years after the invention of the Clovis point, it has sometimes been claimed that this weapon was so effective in the hands of early hunters that it alone led to the extinction of mammoths, mastodons, the ground sloth, and a host of other animals that had flourished during the Pleistocene. More rational heads, however, have argued that the Clovis point was but one of the factors that led to the massive extinctions, with the changing post-glacial climate perhaps playing a more central role.

Many scholars eventually came to believe that no one *could* have lived in the New World before the invention of the Clovis point, and that no stone tools claimed to be earlier could possibly be manmade. This rigid attitude led to the establishment of what has in recent years been called the "Clovis Barrier," the position that there were no people living in the Americas before the invention of the Clovis point. Most of the pre-Clovis sites were seen as having fatal errors in dating or stratigraphy and could be rejected fairly easily. This, coupled with passionate lobbying by individuals in the "Clovis First" camp, caused the Clovis Barrier to become accepted dogma for a majority of archeologists.

However, through diligence and the good fortune of additional discoveries, those working on earlier time levels were forced to become more rigorous in their techniques, recognizing that the acceptance of pre-Clovis sites in the Americas would require the highest professional standards and meticulous record-keeping and dating. Thus the site of Meadowcroft Rockshelter near Pittsburgh, Pennsylvania (Adovasio and Page 2002), eventually became accepted as the oldest known site in the northeastern United States, and Monte Verde in southern Chile (Dillehay 1989, 1997) ascended to the status of the oldest, almost universally accepted site in the New World, with an age of about 14,500 years. (The band of hunter-gatherers that lived at Monte Verde occupied a long, oval hide tent and lived at that site year-round. The preservation at Monte Verde is so outstanding that the site even contained three human footprints and chunks of uneaten mastodon meat!) Since then, many more archeologists have entered the fray, and now such sites as Topper in South Carolina and Cactus Hill in Virginia have also been proposed as pre-Clovis, pushing man's entry into the New World even further back. Interestingly enough, new theories are postulating a North Atlantic route into the New World, and some scholars believe that peoples from western Europe hugged the Atlantic Coast and traveled down the Eastern Seaboard of the Americas, bringing with them Solutrean-style projectile points from Europe. It has become a truly exciting and dynamic time to be conducting research into New World origins, even if some of these theories will no doubt be rejected in time.

Added to this is the issue of racial origins. When the skeleton of what became known as "Kennewick Man" was discovered on the Columbia River near Kennewick, Washington, in 1996 and was dated to about 9,500 years ago, a particularly lively debate ensued as to whether some of the First Americans might actually have been Caucasian (Chatters 2001; Thomas 2002). In the end this theory was largely rejected, but it helped to raise awareness among the general public of some of the more complicated issues being addressed by scholars working on this early time level.

Even the work conducted at Clovis-era sites has been rejuvenated by the pre-Clovis debate, and excavations since 1998 at the Gault site in Texas have revealed the richest, most longterm Clovis occupation yet discovered in the Americas (Poole 2001; Hadingham 2004). The Gault site director, Michael Collins of the University of Texas at Austin, has enlisted huge numbers of students and volunteers and discovered hundreds of thousands of chert flakes and tools (although only relatively modest numbers of Clovis points). Some of these implements have shown evidence of polish, revealing that some of the tools had been used for cutting plants. Many of the bones excavated from this Clovis site are those of turtles, birds, and small mammals, demonstrating what had long been assumed—that the First Americans were much more likely to eat small game than mammoths! Collins estimates that the Gault site dates from about 12,900 to 13,200 years ago.

The original, relatively simple paradigm had a few bands of Ice Age hunters crossing through Beringia, quickly inventing the Clovis point upon arrival, and spreading across all of North and South America within about a thousand years. This is now widely recognized as being overly simplistic or just plain wrong. The most recent discoveries have created a very charged and divisive atmosphere in which it is almost impossible to say which theories will ultimately be accepted by mainstream archeologists. But increasingly, a great many archeologists now believe that the First Americans came from Asia, perhaps in multiple "waves," and possibly from Europe as well, and that these migrations began at least twenty thousand years ago. The First Americans then spread out and exploited a great many different resources prior to the subsequent invention of the Clovis point.

Paleo-Environment in the Northeast

At the close of the Wisconsin glacial sequence, a park tundra environment existed along the fringes of the ice. Plants growing near the ice front, although sparse, would have included low growing grasses, lichens, and mosses. This environment would have had sufficient resources to support large game animals such as mastodon, bear, beaver, musk ox, elk, bison, horse, and caribou (fig. 1.1).

The park tundra ecosystem eventually gave way to a spruce zone char-

FIG. 1.1. Caribou, an important food source in the Northeast at the close of the Pleistocene period. (The photograph was taken at Nulliak in northern Labrador.) Courtesy of Stephen Loring.

acterized by white spruce, balsam, fir, jack pine, paper birch, and aspen, indicating that the cold glacial climate had turned cooler and wetter. White pine began to dominate the forests in southwestern New England, and the climate was warm enough to support a variety of deciduous trees. This resulted in the growth of small, localized boreal forests. The transition to an oak-hemlock forest occurred by 7000 B.C., and the plant species were increasingly dominated by red oak, hemlock, hickory, and chestnut.

We believe that Paleo-Indians entered the Northeast as the glaciers were retreating during the late Wisconsin period, from 10,000 to 8000 B.C. The topography and vegetation during this period differed considerably from current conditions. While regions of the continental shelf were exposed as a result of the accumulation of available water within the large ice lobes, portions of the interior were inundated between 11,000 and 9,000 B.C. by Lake Iroquois, Lake Albany, Lake Vermont, the Champlain Sea, Glacial Lake Newbury, and other bodies of water. The existence of these lakes precluded early Native habitation in many locations that are now well-suited to human settlement. As a result, evidence of Paleo-Indian occupation and use may be anticipated only on the higher grounds surrounding these embayments.

While extensive glacial activity resulted initially in the development of a tundra-like environment that was relatively unfavorable for extensive occupation, Paleo-Indians followed migrating herds along the retreating ice fronts. Traveling in small bands, these hunters appear to have moved seasonally along the major river valleys, keeping primarily to the elevated terraces, because water levels were higher in the impounded river valleys. Subsistence patterns during this stage revolved around hunting activities, because the spruce-tundra environment of grasses and sedges could only support a hunting-gathering society.

Habitation sites during this hunting-gathering phase would have consisted mainly of small open-air camps, specialized resource procurement sites, and kill sites. Artifacts associated with Paleo-Indian sites in the Northeast include the Clovis point, possibly other styles of fluted points, bifaces, unifaces (such as end scrapers), side scrapers and retouched knives, gravers, pulping planes, and *pieces esquillees*. In addition, hammerstones, anvilstones, and abradingstones have been found on Paleo-Indian sites in the Northeast. Although a basic pattern of culture prevailed throughout this period of time, temporal and regional variations of artifact typology did occur.

Many of the artifacts associated with this period have been isolated finds recovered by local relic collectors, not the result of systematic archeological investigations. Still, similarities in site selection and distribution have demonstrated a preference for major drainage valleys, well-elevated landforms, and regions associated with large wetlands and lakes. The preference for these locations may be somewhat explained by the subsistence patterns of these people. Well-elevated landforms would provide opportunistic vantage points for tracking and hunting the migrating herds of Pleistocene megafauna, upon which these populations relied for part of their subsistence. These browsing and grazing mammals would tend to move through the aligned valley floors and cluster around large bodies of water.

Paleo-Indian Sites in the Northeast

Until fairly recently, few well-preserved Paleo-Indian sites had been identified in New York and New England, but over the past 20 years there has been an explosion in research, and quite a few carefully dug sites now make it possible to explain differences between the Early and Late Paleo-Indian periods, regional variations, resource procurement strategies, and changing tool technologies in ways that were not previously possible. A few generations ago, the Bull Brook site in Massachusetts was easily the most diagnostic Paleo-Indian site in New England (Jordan 1960; Grimes et. al. 1984), along with the Reagen site in Vermont (Ritchie 1957), West Athens Hill and Dutchess Quarry Caves in New York State (Ritchie 1969; Ritchie and Funk 1973; Funk and Steadman 1994) and the Shoop site in Pennsylvania (Witthoft 1952). Today there are well-preserved, carefully dug Paleo-Indian sites in every part of the Northeast, including 6LF21 in Connecticut (Moeller 1980), the Vail site in northwest Maine (Gramly 1982, 1984), the Adkins site in Maine (Gramly 1988), and the Michaud site in Maine (Spiess and Wilson 1987). Enough research has been done to finally propose a series of phases within the Paleo-Indian period based on changes in projectile point morphology: the Bull Brook phase (10,800–10,500 before present [or 1950]), the Michaud-Neponset phase (about 10,200 B.P.), the Crowfield phase (unknown dates), and the Nicholas phase (10,100–10,050 B.P.) (Boisvert 2003–2004: 24–25).

As an example of how extensive and complex some Paleo-Indian sites can be, one of the largest and most interesting in the Northeast is West Athens Hill, located in the town of Athens in Greene County, New York. The site straddles the summit of the hill, and a total of 1,493 recognizable artifacts were recovered during the excavations conducted from 1963 through 1966 (Ritchie and Funk 1973: 16; Funk 2004). These artifacts consisted of bifaces, fluted points (generally composed of Normanskill chert), unifaces (end scrapers, side scrapers, and retouched flake knives), retouched flakes, gravers, and *pieces esquillees*; the stone implements also included pebble hammerstones, hammerstones, and anvilstones. This Paleo-Indian site is almost unique in that it is also a quarry workshop and habitation site. The lithic source (the ore) at this site was quarried by means of the excavation of conical pits to expose the chert-bearing strata. Once removed, the materials were reduced and worked within the approximately two-acre site. The initial occupation and use of West Athens Hill dates from approximately 9000 to 8000 B.C., and Native peoples clearly kept returning there to obtain the fine-quality chert. During a period of time when most settlements were simple, brief campsites, West Athens Hill demonstrates that Paleo-Indians were quite capable of creating sizeable and specialized living and working areas.

The First People of New Hampshire

Before the 1970s, the Paleo-Indian period in New Hampshire appeared to be marked by only a very light human occupation that was represented by the handful of fluted points found throughout the state. It was thought that sites lacked good stratigraphy; no early radiocarbon dates existed; and artifacts were simply not there in appreciable numbers. Unfortunately, there also was a very strong tendency to look just for fluted points, ignoring other tool types that might suggest a Paleo-Indian presence. (More recently, a broader range of implements has been used to demonstrate a Paleo-Indian presence, and these include unifacial tools and "channel flakes," which are produced when manufacturing a parallel-flaked point.)

The scattered finds of fluted points in New Hampshire now include the tip of a jasper fluted point found at the Neville site in Manchester (fig. 1.2; Dincauze 1976, Curran 1994: 50); a complete buff-grey point from the Smyth site in Manchester (fig. 1.2); a red chert fluted point found in Intervale in the upper Androscoggin River Valley (fig. 1.3; Smithsonian Museum USNM #149926; Sargent and Ledoux 1973: 67, Curran 1994: 50); a rhyolite fluted point from the Copeland Collection that was found "near the outlet of Ossipee Lake" (Sargent and Ledoux 1973: 67, Curran 1994: 50); a red chert fluted point from the shore of Massabesic Lake in Auburn (fig. 1.2; Sargent and Ledoux 1973: 67, Curran 1994: 50); and a cryptocrystalline, dark gray/black fluted point found in the Piscataguog drainage east of Manchester that Richard Boisvert considers to be similar to those found at the Vail site in Maine (Boisvert 1994a: 6–7). To this very short list of fluted points may be added a few nonfluted points from the Late Paleo-Indian period, with attributes that are rather similar to the Eden point (best known in the western United

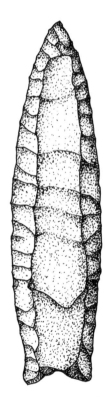

FIG. 1.2. Fluted points found in New Hampshire: (a) Massabesic Lake in Auburn; (b) Smyth site in Manchester; (c) Neville site in Manchester. Photographs by Eugene Finch. Courtesy of Davis W. Finch, Jean F. Topping, and Mary Lou Curran.

FIG. 1.3. The Intervale fluted point. Drawing by Ellen Pawelczak.

States); one was collected by Clyde Berry on the Merrimack River (Bunker 1994: 21) and another by Solon Colby (illustrated here in appendix 4), also probably on the Merrimack River.

An early discovery of a Paleo-Indian site in New Hampshire, within a documented archeological context, was made by Howard Sargent as he excavated a campsite in Newbury. The year was 1952, and Sargent was working at the south end of Lake Sunapee when he discovered a flake scraper and flakes of exotic chert that he considered to be Paleo-Indian (Sargent 1982, 2003–2004). Much later, working at the Russell's Inn site in George's Mills at the north end of Lake Sunapee, Sargent found additional tools—a scraper and a scraper/graver—that he believed to be Paleo-Indian. In 1982 he uncovered the base of a fluted point, a side scraper, and flakes at the same site. Lake Sunapee, as the remnant of a former glacial lake (Glacial Lake Newbury), clearly has early sites on the terraces around its perimeter, and no doubt additional survey work has the potential to locate more sites there.

However, isolated find spots really do not have the potential to reveal very much about the nature of Paleo-Indian life in New Hampshire, and it was not until research was conducted by Mary Lou Curran at the Whipple site in the Ashuelot River Valley in Swanzey that a substantial Paleo-Indian site was located with enough in situ artifacts to be capable of revealing a broad range of activities. The Whipple site was discovered by Arthur Whipple in 1973, and it was offered to Curran in 1975 as a possible dissertation research project at the University of Massachusetts, Amherst. Her excavations began in 1976 and continued in 1977 (figs. 1.4 and 1.5); in 1978 she conducted survey work to check for other sites in the project area. What then followed were years of meticulous analysis and thorough publication that really helped to establish a

FIG. 1.4. The Whipple site: 1977 field season workers. Courtesy of Mary Lou Curran.

FIG. 1.5. The Whipple site: field workers. Courtesy of Mary Lou Curran.

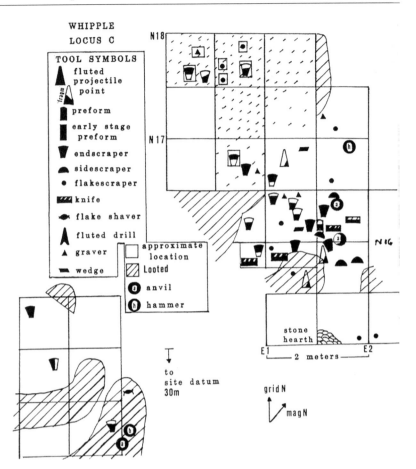

FIG. 1.6. The Whipple site: the tool and tool fragment distribution in locus C. Courtesy of Mary Lou Curran.

FIG. 1.7. The Whipple site: a quartzite fluted point in situ in locus A. Courtesy of Mary Lou Curran.

baseline for Paleo-Indian settlement in New Hampshire (Curran 1980, 1984, 1987, 1994). I visited the excavation several times in 1976 and 1977 and was always impressed by the incredible precision and attention to detail in all of Mary Lou's work.

Three activity loci were identified at the Whipple site, loci A, B, and C (fig. 1.6). The assemblage of stone tools included fluted points, preforms, side scrapers, end scrapers, gravers, flake scrapers, retouched flakes, and a spokeshave, and there was a sizeable quantity of lithic debris (figs. 1.7 and 1.8). The dominant raw material at the site was a grey-black chert that was most likely obtained from a quarry source in Vermont. There were also many bone fragments, including three that were definitely caribou (*Rangifer tarandus*) (Curran 1994; Spiess, Curran, and Grimes 1984–1985: 147–148). Some sixteen radiocarbon dates were obtained from charcoal samples in locus C (from a concentration of bone and debitage), revealing the earliest dates ever obtained

FIG. 1.8. The Whipple site: examples of fluted point fragments. Courtesy of Mary Lou Curran.

for an archeological site in New Hampshire (Curran 1994: 30, Gengras 1996: 10). The oldest of the dates were more than eleven thousand years before present (11,600 +/- 500 years, 11,430 +/- 395 years, and 11,200 +/- 500 years). When these dates are considered along with evidence for hearth areas in loci A, B, and C, the Whipple site will no doubt be viewed as the most informative New Hampshire Paleo-Indian site for a long time to come.

At the same time that fieldwork was being conducted at the Whipple site, Charles Bolian was directing a field school for the University of New Hampshire at Weirs Beach on Lake Winnepesaukee (fig. 1.9). Between 1976 and 1979 Bolian conducted extensive excavations on the beach—one of the most beautiful settings that college students have ever been asked to dig!—and found Archaic occupations underlain by what he considers to be a Late Paleo-Indian component (Bolian 1976–1977, 1980, Maymon and Bolian 1992). At the very bottom of the sequence, atop basal sand, Bolian exposed a hearth that dated to 9615 +/- 225 years B.P. (GX-4569), and next to it was a grey chert unfluted

FIG. 1.9. The excavation at Weirs Beach in 1977. Anne Giesecke is troweling along the base of the wall profile.

biface preform. It is not altogether clear whether this represents very late Paleo-Indian or very early Early Archaic, so it is perhaps best to view this site as right on the border between the two time periods.

Another very late Paleo-Indian site was discovered in 1992 by a SCRAP summer field school headed by then–Deputy State Archaeologist Richard Boisvert. This is the Thorne site (27-CA-26), a small site in the town of Effingham, which is on the boundary between New Hampshire and Maine. As field work proceeded, the team discovered the base of a projectile point, several other bifaces, a possible hammerstone, and debitage that Boisvert believes dates to the very end of the Late Paleo-Indian period, just before the transition to an Archaic lifestyle (Boisvert 2003–2004). The projectile point base was extremely weathered, which makes the raw material identification less certain, although Boisvert believes that it may have been made from a flow-banded spherulitic rhyolite (2003–2004: 34).

A very early, well-preserved Paleo-Indian site was discovered by Victoria Bunker, Inc., in 1998 on an outwash terrace in the Upper Valley of the Connecticut River in Colebrook, New Hampshire (27-CO-38). In a very small area (no more than two by three meters) archeologists discovered what may have been a small fire pit 50 to 63 centimeters below the surface, containing enough charcoal to provide a radiocarbon date of 10,290 +/- 170 years B.P. (the oldest date thus far for human occupation in northern New Hampshire). Twelve small circular stains were found in association with this feature, suggesting to the excavators the possibility of "stakes placed near the edge of a hearth" (Bunker and Potter 1999: 77). Several interpretations may be suggested, including evidence for a structure or windbreak that might have been occupied by a small family, although the authors are cautious about speculating. The associated artifact assemblage consists of the base of a biface fragment of black chert and 325 flakes of dark chert, gray chert, rhyolite, and quartzite. Eight of the flakes are channel flakes, and the debitage "represents several stages of stone reduction, including biface thinning, retouch, and production of cores or rough stone tools" (ibid.). While this is clearly a small Paleo-Indian site with a small artifact assemblage, the integrity of the feature and the artifacts around it has preserved "one moment in time" from a very long time ago indeed. This is also a rare example of a Paleo-Indian site that has two Middle Archaic components above it, demonstrating that Archaic peoples did, in fact, sometimes make the same site selections as Paleo-Indians. The Colebrook site was discovered during field work for the Portland Natural Gas Transmission System, so this is yet another example of the impressive finds that are being made during cultural resource management surveys in New Hampshire.

About 50 miles to the south of the Colebrook site, a series of walkovers was conducted by Edward Bouras and Paul Bock in June of 1995. Their discovery of lithic debitage, including the base of a fluted point in November of that year, led to a series of SCRAP excavations that began in 1996 (Bouras and Bock 1997). Thus far, excavations in Jefferson have revealed a cluster of five Paleo-Indian sites, now referred to as the "Israel River Complex" (and as Jefferson I, II, III, IV, and V). This is the most extensive work since Curran's project at the Whipple site, and it has been conducted under the direction of State Archaeologist Richard Boisvert and the Division of Historical Resources (Boisvert 1998, 2004). Excavations in 1996 revealed three fluted points or point fragments (fig. 1.10), channel flakes, scrapers, retouched flakes, a *piece esquillee*, and exotic raw materials (Munsungun chert from Maine and other, unidentified chert) as well as local rhyolite debitage. The resumption of excavations in 1997 at Jefferson II (27-CO-29) recovered another five fluted point fragments, along with broken bifaces, unifacial cutting and scraping tools, and much debitage (Boisvert 1998: 104–105). Taken as a whole, the Israel River Complex in Jefferson has produced a significant lithic assemblage, evidence for exotic lithics, and no mixing with artifacts from later time periods.

Other recent Paleo-Indian research has included work at the Thornton's Ferry sites (27-HB-1 and 27-HB-2) in Merrimack, New Hampshire. Wesley Stinson originally discovered these sites in 1989 while directing a SCRAP survey, and he identified them as possibly contain-

FIG. 1.10. A fluted point base in situ in Jefferson, New Hampshire, at the Jefferson I site (27-CO-28). Courtesy of Richard Boisvert, NHDHR.

FIG. 1.11. Fluted point fragments from the Thornton's Ferry site (27-HB-2; left) and the Whipple site (27-CH-27). Courtesy of Richard Boisvert, NHDHR, and Wesley Stinson.

ing Paleo-Indian material (Stinson 1991: 7). A Paleo-Indian point base made of felsite was discovered at 27-HB-2, with fluting on one side, though no points were found at 27-HB-1 (fig. 1.11). Richard Boisvert subsequently directed excavations at 27-HB-1 in 2001 and found unifacial side scrapers as well as chert flakes that would have been produced during the manufacture of a parallel-flaked biface (a point style diagnostic of the Late Paleo-Indian period; Boisvert and Bennett 2004). Their interpretation is that "there was manufacture of parallel-flaked bifaces, probably parallel-sided projectile points," and further that "it is possible to isolate Late Paleoindian components on the basis of debitage analysis" (Boisvert and Bennett 2004: 98).

What this means is that while projectile points alone were once used in the identification of Paleo-Indian sites, an increasingly broad variety of other tools, channel flakes, and debitage is now being used to identify the earliest sites in New Hampshire. The number of these sites, their integrity, their evidence for exotic lithic materials, and the extreme precision used in excavating them all bode well for the future of Paleo-Indian studies in New Hampshire.

Chapter 2

The Archaic Period
A Time of Settling In

The Eastern Archaic

WITH THE CLOSE of the Paleo-Indian period, cultures began to make use of a wider range of resources, including more diverse plant and animal species. Still, patterns of movement were becoming more restricted, because increasing population made it almost impossible to move into new regions. What archeologists call "restricted wandering" no doubt led to seasonal rounds and the increased use of locally available raw materials. This means that bands of Native Americans returned to the same spots year after year, taking advantage of natural resources as they became available.

We use the term "Archaic" to identify the long period of time (about 8000–1000 B.C.) that followed the Paleo-Indian, and it is divided into the "Desert Archaic" west of the Mississippi and the "Eastern Archaic" in eastern North America. These two patterns represent adaptations to different environments and resources; the Desert Archaic reflected a time of nonagricultural lifestyle over much of western North America, whereas the Eastern Archaic was the block of time between the earlier Paleo-Indian cultures and the later "Woodland" cultures that had ceramics and horticulture (Thomas 1994: 49–50). The Eastern Archaic left behind literally thousands of small sites that demonstrate how people are able to adapt in myriad ways to lake, forest, riverine, and coastal environments.

It was originally believed that cultures changed very little over the seven thousand years of the Archaic, and Joseph Caldwell at the Illinois State Museum (1958) once argued that Native peoples were so well adapted to their environment(s) that they really didn't need to change. Caldwell called this pattern "Forest Efficiency." However, after many more years of scholarship, it is perhaps more accurate to say that certain broad patterns continued, even as local cultural variations became much more pronounced.

There was one major technological innovation in the Archaic period, and that was the spear thrower, or "atlatl," a mechanical extension of the thrower's arm that allowed hunters to throw their spears farther and more accurately. Caribou, moose, and deer replaced the Pleistocene megafauna and were hunted accordingly, but Archaic peoples also made increasing use of fish and shellfish.

In the East by 6000 B.C. the park-tundra habitat from the end of the Holocene period had been succeeded by a spruce or boreal woodland environment, followed in turn by a temperate deciduous forest with a predominance of oak. This resulted in a diversification of floral and faunal species, and the expansion of these forest species into new territories resulted in an increase in their utilization by prehistoric populations. As a result, social and economic changes occurred within these groups that can be recognized in the archeological record. This evolution of subsistence and technology helps to define the Archaic, a time when hunting and foraging supported steadily increasing populations, and when limited plant cultivation began to appear in some parts of the country. It was the more widespread adoption of farming that finally brought an end to the Archaic lifestyle, although some peoples had not become farmers even by the time of European contact.

The Archaic Period in the Northeast
(Early, Middle, and Late)

The Archaic period was first defined in the 1930s in New York State by William Ritchie, then New York State Archaeologist. Ritchie was working at a site on Lamoka Lake in central New York, and the site showed evidence of a culture that depended on hunting, fishing, and the gathering of wild plant foods. The Lamoka Lake site was also clearly preceramic and preagricultural (Ritchie 1932). Ritchie realized later that this site and many of the sites he subsequently excavated in New York State were representative of a broader stage of economic development, and he applied the term "Archaic" to this pattern of small family groups regulated by a simple social structure (Ritchie 1940, 1969).

For Ritchie, the Archaic also included a large variety of chipped stone tool types, the lack of polished stone artifacts (except for atlatl weights), the use of a large number of bone tools and some copper tools, an absence of shell artifacts and pipes, an absence of pottery (except at the very end of the period), a lack of horticulture, and the presence of a large variety of burial practices (Ritchie and Funk 1973: 37). The Archaic level of culture was later identified all over the eastern United States.

The Early Archaic is typically dated between 8000 and 6000 B.C., and it has been well documented at many sites in the southeastern United

States (see Chapman 1975 and 1977, Coe 1964, and Broyles 1971). In the Northeast, the population appears to have been more modest at this time than it was in the South, although sites have been found on Staten Island (Ritchie and Funk 1971 and 1973: 37–40), on the Taunton River in southeastern Massachusetts (Robbins 1980; Taylor 1976), at the southern end of Lake George in upstate New York (Snow 1977; Starbuck 2002), at the John's Bridge site in Vermont (Thomas and Robinson 1983; Thomas 1992), on the coast of Maine (Petersen and Putnam 1992; Robinson 1992; Sanger, Belcher and Kellogg 1992), in central Connecticut (Starbuck 1980b, 1991), and elsewhere (Dincauze and Mulholland 1977). In the Northeast many isolated finds have also been made of projectile points that have similarities to the Hardaway, Kirk Stemmed, Kirk Corner-Notched, Palmer, and bifurcate-base types (LeCroy, St. Albans, Kanawha) that have been much better defined in the Southeast. However, without a well-dated stratigraphic context for these points, most Northeastern archeologists have been cautious in using these terms. After all, it is very possible that some spear points in the Northeast may be holdovers (survivals) into the Middle Archaic period (Snow 1980: 160–161).

It is hard to say whether the Early Archaic population in the Northeast was much greater than it was during the Paleo-Indian period, and the same was once thought to be true for the Middle Archaic (6000–4000 B.C.). However, excavations at the Neville site in Manchester, New Hampshire, in the late 1960s revealed a well-stratified Middle Archaic sequence (Dincauze 1976), and a host of other excavations since then has demonstrated significant Middle Archaic populations all over the Northeast (Starbuck and Bolian 1980; Robinson, Petersen and Robinson 1992).

By the Late Archaic Period (4000–1300 B.C.) the population density in the Northeast appears to have increased markedly as a result of the establishment of mixed deciduous forests with higher carrying capacities (Ritchie and Funk 1973: 46). The various topographic and environmental settings of Archaic sites suggest the ability of Native peoples to adapt to their changing environments and to harvest a variety of natural resources. Archaic lithic assemblages associated with occupation areas may include oval or trianguloid knives, drills, broad side-notched and corner-notched points, narrow side-notched points, pitted stones, anvilstones, abradingstones, hammerstones, netsinkers, bannerstones, ground stones, adzes, mortars, pestles, and choppers.

Because Archaic sites are often found along small lakes or the shallow portions of large lakes, rivers, streams, or marshes, it can be assumed that a fishing subsistence was important. Remains of fish, along with white-tailed deer, black bear, elk, raccoon, woodchuck, turkey, passenger

FIG. 2.1. Plummets (netsinkers) from the Colby Collection.

FIG. 2.2. Ground slate ulus
from the Colby Collection.

pigeon, and wild plants, have been documented on sites of this period. Acorns, hickory nuts, and butternuts were also harvested during the fall and incorporated into the diets of Archaic populations.

Accumulating data indicate the widespread intensive occupation of Late Archaic groups throughout New York and New England, and archeologists have divided these sites into several traditions, reflecting different cultures or at least different groups of people who were migrating through the region, carrying their own distinctive lifestyles and technology with them. Among these traditions was the Laurentian, first defined by Ritchie in northern New York and Vermont (1969); its diagnostic artifact assemblages include gouges, adzes, plummets (fig. 2.1), ground slate points, knives (ulus) (fig. 2.2), bannerstones, and broad-bladed and side-notched projectile points. This cultural tradition revolved around hunting and fishing, with little evidence of food processing (Ritchie and Funk 1973: 340).

Elsewhere, the Maritime Archaic tradition predominated on the coast of Maine and in the maritime provinces of Canada and was marked by the heavy use of coastal resources (such as swordfishing) and the use of ground slate points. The Small Point or Narrow Point tradition was

FIG. 2.3. Part of a steatite bowl from the Colby Collection.

found across much of New England and New York and was marked by people using Small Stemmed points and Squibnocket Triangle points as well as relying heavily upon quartz for tool manufacture. The last major tradition—and the latest in time—was the Susquehanna, originating in eastern Pennsylvania and marked by the manufacture of broad-bladed points and the use of stone bowls made of steatite (fig. 2.3).

All of these traditions are evidenced archeologically to various degrees throughout New England and New York, often overlapping in the same areas, but it is the Susquehanna that gives the strongest sense of having moved into the region and displaced earlier cultures. (The distinctions among these four traditions are described very effectively in Goodby 2001.)

The Archaic Period in New Hampshire

Whereas the Paleo-Indian period in New Hampshire has only recently received systematic treatment, no such inattention has bedeviled sites of the Archaic period. Most of the early digs by the NHAS were at Archaic sites, as were the digs during the early days of SCRAP (beginning in 1981). Even today most of the sites encountered during cultural resource management surveys date to the Archaic period. This is to be expected given the length of the Archaic period (7,000 years), but site frequency also reflects campsites and manufacturing sites that often were occupied for no more than a day or two, which would have left a tremendous number of small, ephemeral sites scattered across the New Hampshire landscape. Populations were also increasing throughout the Archaic, resulting in growing numbers of archeological sites; by the Late Archaic, the population had reached a level that was probably not equaled again until 3,000 years later in the Late Woodland period. The sequence of development in the Early, Middle, and Late Archaic periods in New Hampshire appears quite similar to that in the other Northeastern states, and it is not until the Late Archaic that there is enough archeological evidence to suggest that a local blending of regional traditions (Laurentian, Maritime Archaic, Small Point, and Susquehanna) might have created a culture that was finally unique to New Hampshire.

In the Early Archaic, New Hampshire had a mix of the same projectile point types that are best known in the Southeast (Kirk, Palmer, bifurcate-base, and others), but while most of these finds are scattered and lack context, there are a few sites that do in fact have stratified Early Archaic components. These are the Weirs Beach site, the Wadleigh Falls site, and deeper layers at the Neville, Smyth, and Eddy sites (all at Amoskeag Falls). During this period, quartz was commonly used for making stone tools; bifacial tools were relatively rare; and artifact assemblages found today will include ground stone rods, full-channeled gouges, unifacial edge tools, cores, and flakes (Bunker 1994: 21).

By the Middle Archaic, fairly sizeable settlements began to appear on waterways and lakes in New Hampshire, seemingly reflecting a growing dependency upon fish. The Neville, Smyth, and Eddy sites in Manchester clearly demonstrate this pattern, as do the Concord sites of Sewall's Falls and Garvin's Falls. Dena Dincauze's work at the Neville site has helped to demonstrate evolving tool complexes throughout the Middle Archaic, beginning with the Neville and Stark complexes and later evolving into the Merrimack complex just before the end of this period. Middle Archaic workshops away from major rivers have also been found and excavated in Belmont and Tilton (Starbuck 1982b; Howe 2000). Quartz continued to be heavily utilized in the Middle Archaic, but volcanic stone was increasingly worked into tools as well (Bunker 1994: 22).

The Late Archaic saw a virtual explosion of settlement in New Hampshire, with many sites being periodically reoccupied as restricted wandering prompted people to return to the same locations. An even greater variety of lithic materials was utilized than before, including quartz, quartzite, chert, hornfels, rhyolite, and crystal quartz, and sites have been found in every possible setting. Most of the sites that had been occupied in the Middle Archaic were reoccupied in the Late Archaic, and fishing stations on the rivers, lake terraces, and places where streams run into rivers were perhaps the most favored locations for settlement.

It is not possible in this book to cover all of the Archaic sites that have been excavated in New Hampshire, but what follows is a selection of some of the more significant or interesting sites. Many had previously been surface-collected or dug by collectors before they were dug by archeologists. They are organized according to region of the state and chronologically within each region whenever possible; regrettably, there is not a section devoted to the Connecticut River Valley, because there is a dearth of useful publications describing Archaic sites on the New Hampshire side of the river. (However, there is an excellent privately printed inventory of prehistoric sites in the Upper Connecticut River Valley—see Cassedy 1991.) Many of the sites described below are multicomponent, but for most of them I plan to focus only upon the time period(s) that is most distinctive or representative.

Early Digs by the New Hampshire Archeological Society

While Warren Moorehead's survey work along the Merrimack River in 1930 included excavations at the Smyth site (1931: 61–64), the first substantial work on the Archaic began with members of the NHAS.

Their first major dig was on Clark's Island in Tilton (beginning in 1949; Sargent 1950, 1951, 1975), followed by society digs at Pickpocket Falls in Exeter (Sargent 1959), the Edgerly Farm in Hampton Falls (a survey in 1957), the Great Bay site in Greenland (dug from 1957 to 1966; Finch 1969), Litchfield Farm in Litchfield (dug from 1959 to 1962 and in 1965), the Garvin's Falls site (dug from 1963 to 1970; Winter 1985), the Smyth site (dug from 1967 to 1968; Winter 1975), the Neville site (dug from 1967 to 1968), and the Stanley site on the Exeter River (dug from 1961 to 1966; Finch 1967). After 1978, the work at the Stanley site, which dates from the Late Archaic to the Early Woodland periods, was continued by Donald Foster and classes from Phillips Exeter Academy (Foster 1982).

All of these sites had Archaic components, often followed by Middle and Late Woodland occupations. The Smyth, Neville, and Garvin's Falls sites will be discussed in detail below, but one of the most interesting of the aforementioned sites is the Litchfield site, on the farm of the Colby Brothers. Here archeologists discovered a Susquehanna broad point that had been "killed," as well as a cremation burial and large numbers of Middle Archaic Stark points (figs. 2.4 and 2.5; Finch 1971).

Many of the NHAS digs were held in conjunction with Howard Sargent, who independently directed many digs throughout the state. The Archeology Club of Phillips Exeter Academy also helped at the Great Bay and Litchfield sites. With the passage of time, the excavation directors sometimes changed, but many successful digs were directed by Eugene Finch, Eugene Winter, and Paul Holmes, and several individuals—notably William "Bill" White—became known for their success in locating significant sites.

FIG. 2.4. A stratigraphic profile at the Litchfield site. Courtesy of the New Hampshire Archeological Society.

FIG. 2.5. Middle Archaic projectile points excavated at the Litchfield site in the early 1960s. Courtesy of the New Hampshire Archeological Society.

The Merrimack Valley

Some of the richest Archaic sites in New Hampshire are located at falls on the major rivers, where fishing and associated activities were conducted over thousands of years, leaving behind substantial middens. Preeminent among these is Amoskeag Falls on the Merrimack River in Manchester, where archeological sites have thick soil layers that are perhaps unequaled anywhere else in New England. As already mentioned,

FIG. 2.6. Bridge construction at the end of the Amoskeag Bridge. The Smyth mansion is in the background. Courtesy of the New Hampshire Archeological Society.

Warren Moorehead dug there (1930, 1931), as did Harlan Marshall (1942) and others. But between 1967 and 1969, plans to replace the Amoskeag Bridge with a larger structure prompted a small, dedicated group of volunteers, working under the auspices of the NHAS, to conduct excavations adjacent to the bridge on the grounds of the historic Smyth mansion, where Governor Frederick Smyth once had lived (fig. 2.6). In addition to the NHAS, which worked under the direction of Eugene Winter, there was a group of students from Franklin Pierce College who worked there in the summer of 1968 under Howard Sargent (Foster, Kenyon, and Nicholas 1981).

A local volunteer, Peter McLane, started a second excavation in the yard of the Neville family. McLane's small group dug through thick deposits of sand, and underneath the expected Woodland deposits they discovered an intact cultural sequence that dated back through the Late Archaic and into the Middle Archaic, a time that had not previously been clearly identified elsewhere in the Northeast. Radiocarbon dates supported this assessment, including 7,740 +/- 280 years B.P. (GX-1746), 7,650 +/- 400 years B.P. (GX-1747), 7,210 +/- 140 years B.P. (GX-1922), and 7,015 +/- 160 years B.P. (GX-1449) on charcoal from the deepest deposits.

The excavation at the Neville site came to an end in September of 1968, once earth moving began for the bridge. Even though most of the site was destroyed at that time, McLane's group had succeeded in attracting the attention of professional archeologists throughout the Northeast. Regrettably, McLane was terminally ill and unable to complete a final report, but he and his colleagues contacted Harvard University and requested that Dena Dincauze analyze the Neville material and write up the report. Fortunately, she agreed (Dincauze 1971, 1975, 1976). The Neville site thus came to represent a very special combination of factors: an unusually deep, well-stratified site; exceptionally rich deposits built up over thousands of years of fishing and resource procurement at the falls; a highly dedicated group of volunteers who made an enormous personal commitment to this site before it was destroyed by bridge construction; and a dedicated scholar who was ready to rigorously analyze and publish on the collection, thereby ensuring that the results would be widely disseminated. McLane's love for archeology and Dincauze's professional analysis helped ensure that the Neville site would become known as one of the richest, most informative Middle Archaic sites in the Northeast.

Some of the projectile points discovered at the Neville site needed to be classified by Dincauze into new types (the Neville, Neville Variant, Stark, and Merrimack points), while others were identified based on existing classifications. The discovery that atlatl weights, full-grooved axes, and possibly ulus and gouges went back in time to the Middle Archaic was significant. For Dincauze, the Neville site represented "a seasonal base camp for a community of people" and "a general-purpose site where a full range of seasonally appropriate activities took place" (Dincauze 1976: 138). No other site in New Hampshire has achieved the importance of the Neville site, and thanks to Dincauze's systematic analysis it has really become the "type site" for the Middle Archaic in the Northeast, against which all other sites are compared.

While work proceeded at the deep Neville site, the dig at the Smyth site was proving to be just as notable, though for other reasons. The Smyth site had been an important fishing station, a village site for the Penacooks, a place of contact between Indians and Europeans, and in later times it was the yard for the home of Governor Frederick Smyth. While deposits may not have been as deep as at Neville (fig. 2.7), and there was not as much material from the Middle Archaic period, the archeologists at the Smyth site in the 1960s dug more than 360 square meters and recovered Neville, Stark, and Merrimack points, while the Late Archaic points included Brewertons, Small Stemmed, Atlantic, Susquehanna Broad, and Orient Fishtail (figs. 2.8–2.10). The Woodland was even better represented than the Archaic, and sherds from hundreds of pottery vessels were found, spanning all of the Woodland period (Kenyon 1981, 1983, 1985b).

FIG. 2.7. A soil profile at the Smyth site. Courtesy of the New Hampshire Archeological Society.

FIG. 2.8. Middle Archaic projectile points recovered from the Smyth site.

With both the Neville and the Smyth sites largely destroyed by the bridge construction, the next major, deeply stratified site to be excavated in Manchester was the Eddy site (NH38–6), downriver and across the falls from Neville and Smyth. Victoria Bunker directed a SCRAP excavation there for nine weeks in 1985, and her team opened 16 square meters. Only limited field records had been kept at Smyth and Neville because of the urgency of the impending bridge construction, so the Eddy site provided the opportunity to have the sort of tightly controlled excavation that really was needed in the area (fig. 2.11; Bunker 1992).

Diagnostic artifacts and radiocarbon dates revealed that the Eddy site spanned the Early, Middle, and Late Archaic periods, as well as the Early and Middle Woodland. Projectile points included Merrimack and Small Stemmed points, and there were numerous scrapers, cores, edge tools, and hammerstones, as well as a sizeable quantity of (deeper) flakes of quartz and (shallower) flakes of "Ossipee" rhyolite. A total of seven radiocarbon dates was obtained: 3,315 +/– 90 years B.P. (GX-12385), 3,680 +/– 80 years B.P. (GX-12386), 3,880 +/– 310 years B.P. (GX-12391), 4,335 +/– 85 years B.P. (GX-12387), 7,595 +/–120 years B.P. (GX-12390), 7,755 +/– 440 years B.P. (GX-12389), and 7,830 +/– 100 years B.P. (GX-12388) (Kenyon 1986: 4).

FIG. 2.9. Late Archaic projectile points recovered from the Smyth site.

FIG. 2.10. Scrapers recovered from the Smyth site.

FIG. 2.11. The Eddy site, showing the removal of a central excavation block. Courtesy of Victoria Bunker.

FIG. 2.12. The Merrimack
River at Garvin's Falls.

FIG. 2.13. Garvin's Falls artifacts in the
Colby Collection. Top left: ovoid edge
tool. Top right: biface implement blade.
Bottom row (left to right): antler awl,
perforated; gouge with contracting butt
and full channel; pendant, perforated;
small biface.

FIG. 2.14. Garvin's Falls, area 3, hearth in pit 4. This hearth radiocarbon-dated to 1465 B.C. and was associated with Wading River and Small Stemmed II projectile points.

North of Manchester, the next major falls on the Merrimack River is Garvin's Falls, at the south end of Concord (figs. 2.12 and 2.13). Archeological sites at Garvin's extend for perhaps a mile or more along the east bank of the river. Excavations were conducted there from 1963 to 1970 and again in 1973 by the NHAS under the direction of Eugene Winter (Winter 1985), and also in 1973 by students from Belknap College under Howard Sargent. Between seasons of digging at Sewall's Falls in Concord, I also led a SCRAP excavation at Garvin's in 1982, conducted through the auspices of the NHAS. Our team discovered a nearly continuous sequence from the Middle Archaic through the Late Woodland periods, and we found that the Middle Woodland was perhaps the best represented. Whereas cultural deposits at the Neville site reached a maximum depth of nearly two meters, those at Garvin's were found to average only one meter (Starbuck 1983b: 28, 1985a). The several hearths that we excavated included one that dated to 3,415 +/– years B.P. (GX-9796) or 1465 B.C. (fig. 2.14). Middle and Late Archaic points, drills, bifaces, and scrapers were all quite common (figs. 2.15–2.17).

Several miles north of Garvin's is the site of Sewall's Falls (fig. 2.18). Sewall's is perhaps best known because it is cited in multiple early

FIG. 2.15. Middle Archaic projectile points recovered from Garvin's Falls.

FIG. 2.16. Drills (perforators) recovered from Garvin's Falls.

FIG. 2.17. Artifacts recovered from Garvin's Falls in 1982. Top row (left to right): plummet (schist); biface fragment (felsite); biface (quartzite); biface fragment (chert). Bottom row (left to right): steep-bitted flake scraper (rhyolite); flake end scraper (porphyry); flake end scraper (porphyry); quartz crystal scraper; quartz crystal scraper.

FIG. 2.18. The Merrimack
River at Sewall's Falls.

sources as the site of one of Passaconaway's forts during much of the
seventeenth century (Lyford 1903; Bouton 1856), and there was even
a trucking house built next to the Indian fort in 1659 (Lyford 1903:
79–80). Also, in the middle of the Merrimack River, the site of Sewall's
Island figures prominently in history as the "principal summer residence" of Passaconaway (Schoolcraft 1851–1857: vol. 5: 230). Considerable collecting was done here by Solon Colby, Lincoln Adams, and
others, especially after the Merrimack River flooded in 1936 and 1938,
causing many artifacts to wash out of the east bank of the river and from
the eastern side of Sewall's Island. These collections suggested a sizeable
Middle and Late Archaic population, in the form of Neville and Stark
points, many Small Stemmed quartz points, gouges, adzes, four grooved
axes, three semilunar slate knives (ulus; see fig. 2.2), a winged atlatl
weight, and more. There was also much Late Woodland pottery found
along the river bank, and Colby found "an infant burial on the east bank,
wrapped in pine bark and eroding out of the river bank" (Howard Sargent, personal communication, 1981).

FIG. 2.19. Excavating rows of shovel test pits at the north end of Gold Star Sod Farm (Sewall's Falls) in the summer of 1981.

FIG. 2.20. Using a Munsell Soil Color Chart at Sewall's Falls. Mary Dupre is on the left and Antonett "Toni" Howe is on the right.

This evidence for a very long occupational sequence, coupled with a research interest in discovering Passaconaway's fort, led me to direct the very first SCRAP excavation on the east bank at Sewall's Falls in the summer of 1981 (figs. 2.19 and 2.20), followed by additional seasons in 1983 and 1984 (Starbuck 1982a, 1983b, 1984b, 1985c). I divided the site into several zones of differing archeological potential, and we then sampled large areas of intervale (the former flood plain of the Merrimack) with systematic shovel test pits. We failed to find evidence for either the fort or the trucking house, but instead we discovered a series of small Late Archaic through Late Woodland sites that probably represented fishing stations at the falls. These were distributed chiefly across what is now the northern end of Gold Star Sod Farm, but we also found sites on one of the bluffs overlooking the falls and at the northeastern corner of Sewall's Island. Evidence generally consisted of stone-lined hearths (figs. 2.21–2.23), accompanied by considerable debitage, some bone, and small numbers of Late Archaic projectile points (fig. 2.24). While deposits were generally very shallow on the east bank of the river (unlike Amoskeag Falls and Garvin's Falls), we nevertheless found literally dozens of hearth features there, and we found deeply buried sites on Sewall's Island as well.

In the summer of 1985 we decided to shift our focus to the west bank of the Merrimack at Sewall's Falls, and we excavated shovel test pits and then one-meter squares both above and below the falls and dam. To

FIG. 2.21. Sewall's Falls, a large "roasting platform" in the sod 9 area on the east bank of the river. This measured 80 × 90 centimeters but lacked clearly associated artifacts, and there was insufficient charcoal for radiocarbon-dating.

FIG. 2.22. Sewall's Falls, a "roasting platform" in pit D6, dated to 3,205 +/− 140 years B.P. (1255 B.C.), thus falling at the very end of the Late Archaic period.

FIG. 2.23. Sewall's Falls, a hearth in sod 6, pit N99E9.5. There were four Late Archaic projectile points (all of different types) in close association with this feature.

FIG. 2.24. Projectile points recovered from Sewall's Falls in 1981.

distinguish it from our previous testing on the east bank, we identified this area as the "Beaver Meadow Brook" site and subsequently found much deeper stratigraphy, as well as a good range of both Middle and Late Archaic artifacts. These included a full-grooved ax (fig. 2.25), Neville points (fig. 2.26), many Small Stemmed quartz points, several drills, scrapers, anvilstones, a complete gouge, lots of quartz and hornfels debitage, hundreds of bone fragments (chiefly turtle), and much more (fig. 2.27). We also found several hearths and evidence for a sizeable Woodland period occupation (see Howe 1988; Starbuck 1985d).

The Lakes Region

The Lakes Region of New Hampshire has been a popular destination for thousands of years, both as a wonderful place to live and as a source of lithic materials suitable for manufacture into tools. The Weirs-Aquadoctan archeological district is located here, and easily the

FIG. 2.25. A full-grooved ax from the Neville assemblage at the Beaver Meadow Brook site.

FIG. 2.26. Neville bifaces at the Beaver Meadow Brook site. All are projectile points, though the specimen at bottom right has been reworked into a drill (perforator).

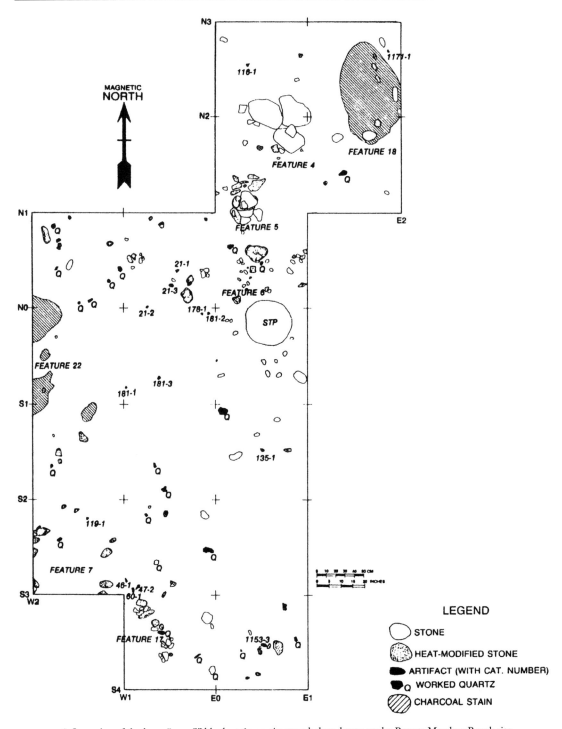

FIG. 2.27. A floor plan of the large "area S" block at 65 centimeters below datum at the Beaver Meadow Brook site. Courtesy of Dennis E. Howe.

FIG. 2.28. A stratigraphic
profile at the Weirs Beach site
in the summer of 1977.

best-known archeological site within the district is Weirs Beach, located
close to the outlet of Lake Winnipesaukee and also very near to the
headwaters of the Merrimack River. Charles Bolian and a University
of New Hampshire crew worked at Weirs Beach (fig. 2.28) from 1976
to 1979 (Bolian 1980; Bolian and Cressey 1978; Maymon and Bolian
1992), whereas Howard Sargent worked nearby at Aquadoctan in 1977
(Sargent 1980).

As already noted, there was an occupation at Weirs Beach at the very
end of the Paleo-Indian period, and a radiocarbon date of 9,615 +/– 225
years B.P. has been obtained for this component. This has recently been
termed Late Paleo-Indian, and it was followed by an Early Archaic component that is marked by "large amounts of quartz debitage, cores, steep-
bitted quartz scrapers and elongate stone rods" (Maymon and Bolian
1992: 118). A total of three stone rods was found at Weirs Beach; two
were found in a pit feature dated to 8,985 +/– 210 years B.P. (GX-4571)
and the third in a pit feature dated to 9,155 +/– 395 years B.P. (GX-5445).
The rods are made of schist, and their use has been the subject of endless speculation with no real solution. Altogether Jeffrey Maymon and
Charles Bolian have located some eight rods in collections that have
derived from Weirs-Aquadoctan (1992: 120). One bifurcate base point
was also excavated by Bolian at Weirs Beach from a relatively late pit
feature radiocarbon-dated to 7,315 +/– years B.P. (GX-4568). At the
time of its discovery, this Early Archaic material was some of the oldest Archaic evidence ever found in northern New England, and it lies

FIG. 2.29. Excavating at the Belmont site (NH31-20-5) in 1981.

FIG. 2.30. Overview of the excavated South Workshop at the Belmont site.

underneath later Neville and Stark points and associated materials from the Middle Archaic. Peoples have therefore lived at Weirs Beach and in the surrounding areas almost continuously from Late Paleo-Indian times up until the present day, and this area has an abundance of natural resources that have always made it a favorite habitation site.

All of the Lakes Region is rich in raw materials for stone tool making, and between April and June of 1981 I directed salvage excavations for the University of New Hampshire at the site of two Middle Archaic lithic workshops located just east of the Winnipesaukee River in Belmont (fig. 2.29; Starbuck 1982b, 1983b). The site (NH31-20-5) is located about 11 kilometers northeast of where the Winnipesaukee flows into the Merrimack at Franklin, and it is on the side of a glacial drumlin about seven to ten meters above the present level of the Winnipesaukee River. By placing one-meter-square test pits at five-meter intervals across the site, we successfully located the two workshop areas—where local rhyolite had been manufactured into tools—and then opened up large blocks of pits to expose nearly all of each workshop (fig. 2.30).

Although we were not able to obtain any radiocarbon dates on the earliest portions of our Belmont site, we found six Neville points, four Stark points, one Neville Variant point, and many perforators, quartz scrapers, preforms, and flake tools, as well as a few hammerstones (figs. 2.31–2.34). The tools were nearly identical to those in strata 5A and 5B at the Neville site (Dincauze 1976), suggesting a date of about 6000–5000 B.C. for the workshops. Because we found large quantities of

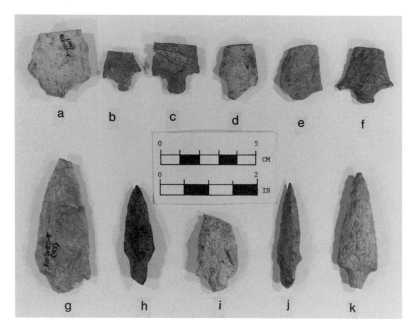

FIG. 2.31. Middle Archaic points recovered from the Belmont site: *b, c, d, f, i, k* are Neville points; *a, e, h, j* are Stark points; *g* is a possible Neville Variant point.

FIG. 2.32. Perforators recovered from the Belmont site: *a, c, d, e, f* are plain; *b* is a perforator shaft on a flake; *g* is a perforator on a Stark point base; *h, i, j, k* are perforator shafts on unspecialized bifaces.

FIG. 2.33. Bifaces of rhyolite recovered from the Belmont site.

FIG. 2.34. Scrapers recovered from the Belmont site: *a* is a Meadowood biface trianguloid end scraper; *b, c* are quartz crystal scrapers; *d* is an expanded-bit scraper; *e, f* are flake end scrapers; *g, h, i, j, k* are steep-bitted flake scrapers.

perforators, scrapers, and gravers, we believe that people may have been processing hides and wood at the site as well, so these really were more than just workshops—people obviously lived here even as they manufactured tools, and they no doubt reduced cores into preforms prior to transporting them on to other camps.

Altogether we dug 111.75 square meters in Belmont, and we then performed a detailed attribute analysis on about 20 percent of the debitage that had been recovered. We recovered 19,162 flakes from the northern workshop and 8,541 from the southern workshop. Afterward, we selected 7,202 flakes out of the total of 29,530 flakes and chips and measured almost every last possible characteristic of them. This sort of detailed analysis would ideally be done with *every* lithic assemblage we recover, but archeologists do not always have the opportunity to do so. And so we spent hundreds of hours determining the raw material, core reduction category (position within the core reduction process), flake length, flake width, flake thickness, relative size, amount of dorsal cortex present, number of dorsal ridges present, depth of the striking platform, width of the striking platform, relative depth of the striking platform, shape of the striking platform, nature of platform preparation (prior to removal of the flake), presence or absence of a lip on the ventral edge of the platform, and character of the bulb of percussion. And this was done for every one of those 7,202 flakes!

It has often been said that archeologists must be incredibly compulsive to "do what we do," but the reality is that this sort of analysis can

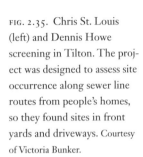

FIG. 2.35. Chris St. Louis (left) and Dennis Howe screening in Tilton. The project was designed to assess site occurrence along sewer line routes from people's homes, so they found sites in front yards and driveways. Courtesy of Victoria Bunker.

help to determine the exact steps whereby stone tool manufacturing is done. We would like to find out how techniques of manufacture changed over time, how this varied with different cultures, and how techniques differed for different types of stone. Lithic analysis is slow, painstaking work, but it is the only way to really understand how prehistoric peoples reduced cores of raw material (stone) into finished tools. The analysis done on our two workshops in Belmont was a great way to determine how Native peoples had reduced rhyolite from outcrops in the vicinity of Lake Winnipesaukee and the Winnipesaukee River into the projectile points and other tools that we find at prehistoric living sites all over New Hampshire. Once all of our measurements had been taken, one of our student analysts even wrote his master's thesis on what we had learned (Cassedy 1984).

The Lakes Region has many good sources of stone that were used by Native peoples, and there are unquestionably a great many other Middle Archaic lithic workshop sites in the vicinity of the Winnipesaukee River and in the larger Lakes Region. In 1999 Victoria Bunker, Inc., excavated two workshops similar to those in Belmont on the property of the Pierce and Fitzgerald families near Silver Lake in Tilton (fig. 2.35). Both were single-episode, single-component (one occupation) sites where the occupants appear to have processed raw material from the Lakes Region and then supplied finished tools (or at least preforms) to populations in the Merrimack Valley and elsewhere.

Both of the sites in Tilton contained much lithic debitage, cores, primary bifaces, preforms, and finished tools, and once again rhyolite was the primary lithic material selected for tool manufacture. At the Pierce site, there were 15 Neville and Neville-like points (or fragments) and two fragments of a single Stark point; there were also 4,191 flakes of rhyolite that were "created during intermediate and final stages of the point manufacturing process" (Howe 2000: 7). The second workshop, on the Fitzgerald property, proved to be very similar, and it contained six Neville and Neville-like points and 8,221 flakes and chunks of rhyolite. Given the strong similarity of manufacturing techniques in the two workshops, Dennis Howe has prepared a flow chart to show the successive stages in the process of manufacturing a Middle Archaic Neville projectile point (fig. 2.36). Howe argues that the consistency of final product (the Neville point) is the result of "a tightly managed process" (2000: 30): "Rather than each Middle Archaic individual having to learn all the skills needed to survive or perish, it is more likely that a highly organized social and economic system assured their survival. The evidence of specialization revealed by the lithic workshops is compelling" (2000: 33).

The richness of sites located between the Lakes Region and the Merrimack River is further exemplified by the Lodge Site (NH31-6-6) in Tilton, which is located on the west bank of the Winnipesaukee River in the Lochmere Archeological District. Work was conducted there under the auspices of SCRAP in the spring and summer of 1984, revealing a multicomponent site spanning the Middle Archaic through the Late Woodland periods. While it does appear that workshops had once existed there, they had been destroyed by bulldozing (Gengras and Bunker 1998: 9). Nevertheless, the Lodge site contained many diagnostic Middle and Late Archaic tools (three Neville points and three Neville Variant points from the Middle Archaic, and Brewerton Eared-Notched, Genessee, Normanskill, Beekman Triangle, Squibnocket Triangle, Wading River, and Orient Fishtail points from the Late Archaic). Many primary bifaces were found, as well as 23 scrapers and scraper fragments, 14 flake tools, six perforator fragments, two gravers, 31 cores, and a fragment from an atlatl weight. Some 3,984 flakes were found at the Lodge site, as well as 145 pottery sherds from the Woodland period. The Lodge site was regrettably similar to a great many other sites in the Lakes Region in that it had very shallow stratigraphy, such that artifacts from all time periods were mixed together, and the site was discovered during clearing for house construction in the fall of 1983, so it needed to be rescued at the last minute under less than ideal conditions. Nevertheless, the Lodge site demonstrates that much useful information may be gleaned even from sites that are shallow and heavily disturbed, and useful comparisons may be made to other sites in the region. A hearth (feature 1) and its associated Genessee-like biface were radiocarbon-dated to the Terminal Late Archaic, 3,840 +/- 70 years B.P. (Beta-62145)

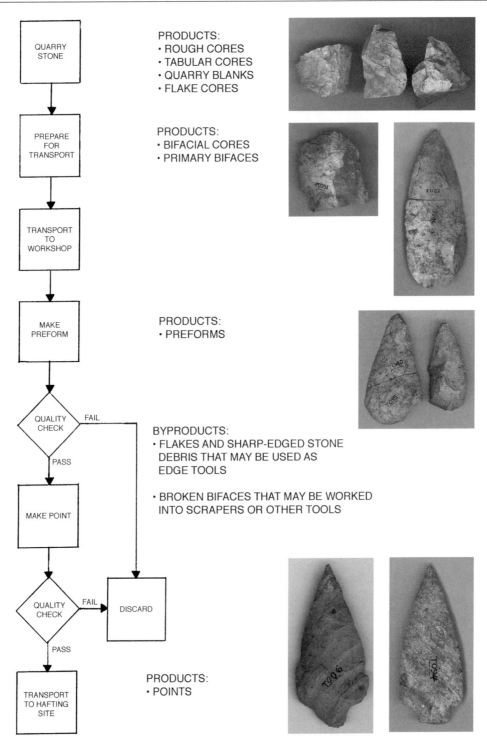

QUARRY
STONE

PRODUCTS:
• ROUGH CORES
• TABULAR CORES
• QUARRY BLANKS
• FLAKE CORES

PREPARE
FOR
TRANSPORT

PRODUCTS:
• BIFACIAL CORES
• PRIMARY BIFACES

TRANSPORT
TO
WORKSHOP

MAKE
PREFORM

PRODUCTS:
• PREFORMS

QUALITY
CHECK   FAIL

PASS

BYPRODUCTS:
• FLAKES AND SHARP-EDGED STONE
  DEBRIS THAT MAY BE USED AS
  EDGE TOOLS

• BROKEN BIFACES THAT MAY BE WORKED
  INTO SCRAPERS OR OTHER TOOLS

MAKE POINT

QUALITY
CHECK   FAIL   DISCARD

PASS

TRANSPORT
TO HAFTING
SITE

PRODUCTS:
• POINTS

FIG. 2.36. A flow chart of the Neville projectile point manufacturing process. Courtesy of Dennis E. Howe.

(Gengras and Bunker 1998: 28), but it was not possible to obtain a radio-carbon date for the Middle Archaic component at the Lodge site.

North of the Lakes Region

The northern portions of New Hampshire have clearly not seen anywhere near as much research or publishing on the Archaic as have the central or southern parts of the state. This is partially, or largely, due to its distance from universities and state agencies that conduct archeological research, but it may also be the result of having fewer of those earthmoving activities that tend to discover sites by accident. However, there are a small number of sites that have seen productive research, and among these is the Mount Jasper Mine (NH27-CO-1) in Berlin, New Hampshire, entered on the National Register of Historic Places in 1992 as the "Mount Jasper Lithic Source" (Boisvert 1992). This prehistoric quarry was, according to Michael Gramly, a "small-scale" source of flow-banded rhyolite for prehistoric peoples and is located just northwest of the confluence of the Dead and Androscoggin Rivers (Gramly 1980: 1). Gramly began excavations at the quarry in 1976, and he found two workshop areas (the Dead River Workshop and the Hill Workshop); he also dug a short distance to the north at Molls Rock on Lake Umbagog (Gramly and Cox 1976).

Gramly found evidence for the various workshops here having been used for thousands of years, and he found sizeable quantities of flakes, cores, small- to medium-sized hammerstones, and projectile point pre-forms "in all stages of manufacture"; the materials here numbered "in some places several thousand specimens per square meter" (1980: 4). The projectile points he recovered, manufactured from Mount Jasper rhyolite, included Middle and Late Archaic types (as well as a Susquehanna Broad point and Middle Archaic points made from other materials), and he found numerous drills, end scrapers, side scrapers, utilized flakes, and assorted other bifaces. Gramly's excavations at Mount Jasper, followed by Richard Boisvert's studies of the site (1992), have yielded considerable information about early mining activities, and this really is one of the most interesting, albeit specialized, sites in the White Mountains.

Some distance to the south, between Lake Winnipesaukee and the White Mountains, is a very important cluster of hornfels quarries in Tamworth, New Hampshire. A project was conducted here along Route 25 for the New Hampshire Department of Public Works and Highways, and the survey located a number of previously unknown sites (NH20-5, NH20-6, NH20-7, NH21-11, NH21-12, and NH21-13) (Ewing and Bolian 1991). While no projectile points were recovered during the

FIG. 2.37. Two large primary bifaces recovered from NH21-12 in Tamworth. Hornfels was used as a raw material in the production of large bifacial cores, primary bifaces, and implement blades. Courtesy of Victoria Bunker.

Tamworth survey, Robert Ewing and Charles Bolian nevertheless found two hornfels bifaces at NH21-12 (fig. 2.37) that they believe probably date to the Late Archaic (1991: 92). Additional, extremely productive work was conducted in the area several years later when then-Deputy State Archaeologist Richard Boisvert directed a SCRAP project at other sites in Tamworth. There is no question that a high density of hornfels quarry sites exists in Tamworth, utilized by Archaic and later peoples, but these sites are extremely sensitive and require protection. A really thorough archeological assessment needs to be done so that they may be preserved for future research.

Southeastern New Hampshire and the Seacoast Region

The most significant known Middle Archaic site in the seacoast area of New Hampshire is located on an island just below Wadleigh Falls on the Lamprey River in Lee. This site (NH39-1) was originally excavated by an amateur archeologist, Gary Magnuson, in 1969 and 1970, and then it was tested again in 1980 by the Coastal Zone Survey of New Hampshire. The 1980 project was conducted by Archaeological Research Services at the University of New Hampshire (Pope 1981; Skinas 1981), and the UNH team used shovel test pits, followed by larger test squares, to sample the site. Subsequently, a field school was held at the site in 1982, directed by Robert Ewing.

As originally described by Skinas (1981) and later elaborated upon by Jeffrey Maymon and Charles Bolian (1992), the UNH team recovered an excellent range of projectile points from the Middle Archaic period, followed by very small quantities of material from the Late Archaic and Woodland. In 1980 the team found six Neville points, two Neville Variant points, and two Stark points, along with considerable flaking debris of felsite, hornfels, rhyolite, quartz, quartzite, and other materials. Dr. Howard Hecker (subsequently of Franklin Pierce College) identified bones from small mammals, turtles, and fish (Skinas 1981: 19), and two hearths were discovered at the site as well.

A single projectile point was recovered at Wadleigh Falls in 1980 that appeared similar to Stanly Stemmed points as found in the Carolina Piedmont (Coe 1964: 35). This left the door open to the possibility that Wadleigh Falls might in time reveal evidence for an Early Archaic component. After the additional research in 1982 and the further updating by Maymon and Bolian (1992), Wadleigh Falls stands out as an exceptionally well-preserved Middle Archaic site, but the exact nature of its earlier occupation is still somewhat unclear. One radiocarbon date was obtained for the lower component, which is dominated by "quartz debitage, cores, unifaces, retouched flakes, and utilized flakes (Maymon and Bolian 1992: 130), and Maymon and Bolian argue that it is thus very similar to the lower component at Weirs Beach. The date yielded by Wadleigh Falls is 8,630 +/- 150 years B.P. (Beta-9050), which is equivalent to 6680 B.C. This certainly falls within the Early Archaic period, but it would be useful to see more diagnostic tools from this lower occupation.

The best representative of the Late Archaic period on the seacoast of New Hampshire is unquestionably the Rocks Road site (NH47-21), formerly known as the Seabrook Station site. A 1973 site survey led to excavations by Charles Bolian and a UNH team in the summers of 1974 and 1975 (Robinson 1983; Robinson and Bolian 1987; Eastman 1974). The UNH team was assisted at various times by members of the NHAS and by students from Phillips Exeter Academy. This was a very sizeable project that was conducted prior to the construction of the Seabrook Station nuclear power plant, and the dig was conducted under a salvage contract between the Public Service Company of New Hampshire and UNH. Over 650 square meters were excavated, revealing periodic occupation from the Late Archaic period up through the early Contact period. Rocks Road was divided into several different sites, and "site 1" was where the most extensive excavations occurred: some 93 features were recorded at site 1, and most of them were shell-filled pits

(Robinson and Bolian 1987: 28). Four human burials were also found at site 1. While a great deal of what was found dated to the Woodland period, there was a modest amount of Late Archaic material, with diagnostic artifacts including two Brewerton projectile points, two Small Stemmed points, and some Atlantic Phase artifacts (in the Susquehanna tradition) that were found in feature O 17-D and that radiocarbon-dated to 3,805 +/- 135 years B.P. (GX-4182) (Robinson and Bolian 1987: 38). The four thousand years at the Rocks Road site were tightly compressed, with Woodland period storage pits confusing the stratigraphy even more, but this proved to be an immensely exciting project for everyone involved, and several of the excavators have subsequently continued to work in the field of archeology.

A short distance from the Rocks Road site, a second excavation was conducted in 1974, also by UNH, at the Seabrook Salt Marsh site (NH47-22). While it may not have gained as much attention as Rocks Road—after all, it's hard to top the visibility associated with a nuclear power plant!—Brian Robinson and some volunteers worked successfully on a partially submerged site that had earlier been discovered and reported by Bill White (Robinson 1976–1977, 1985). Because the tidal marsh has been steadily encroaching on the mainland for thousands of years, this site had to be dug between tides, and "excavation was only possible at low tide for a maximum of three to four hours per day" (Robinson 1976–1977: 1). This was unquestionably one of the most difficult excavations ever conducted in New Hampshire!

Radiocarbon dates obtained for the Seabrook Salt Marsh site fell at 3,720 +/- 150 years B.P. (GX-3663), 3,610 +/- 120 years B.P. (GX-3662), 3,410 +/- 150 years B.P. (GX-3660), and 2,215 +/–125 years B.P. (GX-3661). A human skeleton was well preserved here (Hecker 1981), as were bone and wood artifacts in general. Robinson discovered Squibnocket Stemmed, Squibnocket Triangle, and Wading River points, and also one Susquehanna Broad point; these, together with the radiocarbon dates, place the site squarely in the later stages of the Late Archaic. Distinctive artifacts included an unfinished bone fishhook, four gouges, plummets, net sinkers, and a five-inch (12.7 cm) bone knife (Robinson 1976–1977: 4–5). Several cultural features were also found five feet below the surface of the marsh, including a fire pit, a refuse pit, and four post molds. This had definitely been dry land before the tidal marsh began slowly to creep its way inland.

Working in the Seabrook Salt Marsh had unexpected benefits because of its excellent preservation of organic materials, but also because it provided evidence for deep-sea fishing and because the remains of a great auk were discovered here—the earliest ever found in New Hampshire. The subsequent writing and analysis went on over quite a few years, and

Brian Robinson's commitment to digging and understanding this site proved to be little short of phenomenal!

The Terminal Archaic and the Transitional Stage

Archeologists often split off the final years of the Archaic (about 1300–1000 B.C.) into what is called the "Transitional stage." This short period features the use of carved soapstone vessels among the Late Archaic cultures, just before the introduction of true ceramics (Ritchie and Funk 1973: 71). The Transitional appears to have had its roots in the Susquehanna drainage basin of eastern Pennsylvania, and William Ritchie and others termed this manifestation over space and time the "Susquehanna tradition," as I mentioned earlier. Broad-bladed points are typically associated with this tradition, and they include the Susquehanna Broad point and the Perkiomen Broad point (fig. 2.38). In addition, narrow "fishtail" projectile points serve as diagnostic indicators of this stage (Ritchie 1969, 1971), as do the Atlantic and Wayland Notched points defined by Dena Dincauze in eastern Massachusetts (Dincauze 1968, 1972).

FIG. 2.38. A Perkiomen Broad point from the Colby Collection.

Outside of New Hampshire, the Transitional has been divided into the Orient and Frost Island phases, the former defined by excavations conducted at a burial site located on Long Island (Ritchie 1959). Cemeteries pertaining to the Susquehanna tradition contain cremation burials accompanied by ritually broken or "killed" artifacts (Goodby 2001: vi). While this phase or culture is indisputably native to Long Island, it has manifestations ranging northward and eastward, and the general artifact assemblage includes soapstone pots, atlatl weights, strikers, Orient Fishtail points, drills, ovate knives, large stemmed knives, and other polished tools.

Other tools associated with the Transitional stage are end scrapers, anvilstones, choppers, pestles, netsinkers, expanded base drills (fig. 2.39), and utilized and retouched flakes. Some inland winter camps were established as part of seasonal rounds, utilizing aquatic and terrestrial resources. While the gathering of wild vegetable food was likely, there are no indications that horticulture was practiced. Cooking was accomplished through the use of stone pots, large beds of stones, or pits.

Within New Hampshire there was relatively little evidence for the Transitional stage until very recently, except for stone bowl fragments at a handful of sites, including the Smyth site (fig. 2.40) and the Litchfield site (Finch 1971). There also has been the occasional presence of Susquehanna Broad projectile points, such as the "killed" Susquehanna Broad point found at the Litchfield site—it had been broken into eight pieces (Finch 1971). At the Neville site, Dincauze identified

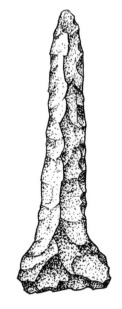

FIG. 2.39. An expanding base drill (perforator) from the Colby Collection. Drawing by Ellen Pawelczak.

FIG. 2.40. Steatite sherds recovered from the Smyth site by the New Hampshire Archeological Society.

FIG. 2.41. Late Archaic bifaces recovered from the Davison Brook site. Courtesy of Victoria Bunker.

an Atlantic spearpoint and knife, ovoid biface scrapers, and a fragment of a winged atlatl weight, all affiliated with the early Susquehanna tradition (1976: 128). She also identified a Susquehanna Broad point, as well as "many typical Susquehanna scrapers, knives, heavy tools, and steatite sherds . . . plucked from the plow zone" (1976: 130).

More recently, two newly excavated sites have provided useful evidence for the very end of the Archaic. The first of these is the Davison Brook site (27-GR-201) in Holderness, where Robert Goodby (2001) recently excavated at the confluence of Davison Brook and the Squam River. Here at locus 1 he discovered large Late Archaic bifaces of the

FIG. 2.42. The excavation at 27-CA-60. Courtesy of Victoria Bunker.

FIG. 2.43. Bifacial cores of hornfels recovered from 27-CA-60. Courtesy of Victoria Bunker.

FIG. 2.44. Intermediate stage primary bifaces recovered from 27-CA-60. Courtesy of Victoria Bunker.

early Susquehanna tradition (fig. 2.41). To the south of locus 1 he also discovered a single sherd from a stone bowl, the exterior of which is "mottled with a layer of carbonized residue, suggesting use as a cooking vessel" (Goodby 2001: 61).

The other terminal Archaic site is 27-CA-60 on the Ossipee River at the Effingham and Freedom town line, where Victoria Bunker conducted a study (fig. 2.42) for the New Hampshire Department of Transportation and found bifacial cores and intermediate-stage primary bifaces that reflect tool making activities in close proximity to a quarry location (Bunker 2002). Site 27-CA-60 lacked any fragments of stone bowls, but Bunker discovered a very short-term occupation that left behind an assemblage dominated by dense deposits of hornfels flaking debris. This was thus a workshop for bifacial tool manufacture (figs. 2.43 and 2.44). She places 27-CA-60 in "the Broad Blade Tradition of northern New England," thus falling between three and four thousand years ago (Bunker 2002: 52).

Chapter 3

The Woodland Period
The Rise of Village Life

The Eastern Woodlands

T HE RISE OF Woodland cultures in the eastern United States was marked by the introduction of pottery, by smoking pipes, by the advent of semi-sedentary village life, and by the increasing use of storage pits and trash pits, horticulture, and mortuary ceremonialism (Ritchie 1969: 179–180). Some of these innovations had their roots in the Late Archaic period, and sometimes these cultural changes traveled together as clusters of traits. Still, there were areas of the country that never adopted horticulture, and there were other peoples that never accepted the elaborate burial customs typically associated with the moundbuilders. It would thus be a mistake to view the cultures of the eastern woodlands as responding uniformly to change.

Given the excellent preservation of the pottery here, the arrival of pottery now may be used as the clearest indicator of cultural movement toward the Woodland way of life. Pottery appeared at the very beginning of the Woodland period over much of the East, whereas every other innovation that has traditionally been associated with the Woodland manifested itself later or, in some cases, not at all. The old model that suggested a new, village way of life based on cultigens that abruptly appeared after thousands of years of Archaic adaptations was never very accurate, and modern archeology suggests that most Woodland cultural changes arrived somewhat gradually, over a lengthy period of time.

The Woodland Period in the Northeast
(Early, Middle, and Late)

The Woodland period in the Northeast (1000 B.C. to around A.D. 1600) is customarily divided into three parts, the Early, Middle, and Late Woodland. The first use of pottery is attributed to the Early Woodland period, but at least initially people's primary settlement and subsistence patterns

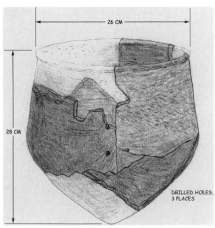

FIG. 3.1. A partially reconstructed Vinette I vessel excavated at the Beaver Meadow Brook site on the west bank of the Merrimack River at Sewall's Falls (Starbuck 1985: 101; Howe 1988: 82–83).

FIG. 3.2. A drawing of the Vinette I vessel in figure 3.1, showing locations of drilled repair holes. The complete vessel would have measured 28 centimeters from base to rim. Drawing by Dennis E. Howe.

remained unchanged. Artifact inventories and food refuse identified on sites associated with the Early Woodland reveal a hunting-fishing-gathering economy with a predilection for large lakes and streams. The typical artifact assemblage includes such items as tubular smoking pipes, distinctive gorgets, birdstones, boatstones, copper ornaments, and bar amulets. Burial ceremonialism also appears to have been a major feature of the Early Woodland period in the Northeast, and burial goods were often made of exotic materials such as Ohio chalcedony, or banded slate, and Harrison County, Indiana, flint (Ritchie and Funk 1973: 96).

The Early Woodland period has been most carefully defined in New York State, where much of the state experienced the Meadowood phase; other states in the region have thus tended to compare themselves to the patterns in New York. Perhaps the most diagnostic characteristic of this phase was the appearance of Vinette I–type pottery, which was originally defined in 1949 as having "complete interior and exterior cord-marking with no decoration, a straight neck, and a conoidal base." In addition, "The entire exterior has been malleated with a cord-wrapped paddle and the cord-markings run in various directions, although they tend to be vertical. Interior surfaces are also completely cord-marked, the impressions always running in a horizontal direction" (Ritchie and MacNeish 1949: 100). Since it was first defined, Vinette I pottery has been found in a great many contexts throughout the Northeast, such that it has become a true time marker for the start of the Woodland (figs. 3.1 and 3.2).

Meadowood peoples have left us with no reliable evidence for the use of horticulture; they are believed to have lived in small bands, perhaps composed of 30 to 50 individuals. Their tool inventory contained birdstones, polished stone and copper plano-convex adzes, knives (made from chert whenever possible), drills, hammerstones, anvilstones, and abradingstones (tabular). Excavations of burials dating from the Meadowood phase indicate that these people were concerned for the well-being of their dead, and burial ceremonialism included cremation and the presence of charnel houses. Often natural glacial mounds were chosen for interment.

In the western parts of the region, there are traits that appear to be shared with the Adena culture of Ohio. This set of traits is identified as the Middlesex phase in New York, and it does have a distribution into Vermont, especially at burial sites such as Boucher in the Champlain Valley (Snow 1980: 291–298). Characteristic artifacts associated with this phase include a tubular pipe form (blocked-end), leaf-shaped lanceolate knives, javelins (spearheads), copper tools, truncated or bust-type birdstones, gorgets, pendants, trianguloid cache blades, copper beads, and discoidal, barrel-shaped cylindrical beads. Widespread burial ceremonialism, including burial mounds and earthworks, is usually associated with this phase as well.

The Middle Woodland period has been defined by the presence of stamped or impressed pottery styles; the types of decoration include dentate, pseudo-scallop-shell, rocker-stamped, and cord-ornamented designs (Ritchie 1969: 180). A distinct Point Peninsula culture has been identified for this period, and Point Peninsula peoples appear to have occupied regions of low relief along the Great Lakes and the St. Lawrence lowlands, in areas heavily forested with hemlock, white pine, and northern hardwoods (sugar maple, beech, and basswood). The limited number of projectile points identified at these sites has suggested to some that hunting was a secondary subsistence activity; fish and freshwater mussels appear to have been the primary source of food. Archeological evidence suggests the extensive use of wild rice beds in some areas, along with the possibility of other cultigens by late in the Middle Woodland. A typical artifact assemblage included simple end scrapers, notched spokeshave scrapers, flake knives, simple bone awls, beaver incisor tools, conical antler and toggle-head harpoons, and compound fishhooks. The mortuary customs were relatively undeveloped compared to those of the Early Woodland period.

Middle Woodland diagnostic artifacts included Fox Creek and Greene projectile points (Ritchie 1971), which lasted until about A.D. 850. Archeological evidence suggests that seasonal rounds were part of the overall subsistence pattern, with people participating in a restricted wandering settlement pattern. The earlier net-marked pottery styles were replaced by cord marking, and corded stick decoration became the prevailing decorative technique after A.D. 700. Also, the triangular Levanna style of point appeared around A.D. 900.

John Hart and Christina Rieth (2002) have recently amassed an impressive amount of scholarship demonstrating that the period from A.D. 700 to 1300 was an especially dynamic time of change, and many of the patterns that we typically associate with the Woodland did not in fact coalesce until then. This is perhaps most true with regard to horticulture, which did not usher in the Woodland, as once thought, but instead was a late and possibly minor supplement to a diverse economy based on both land and aquatic resources. While there is evidence for Woodland Indians in some parts of the country cultivating goosefoot, marsh elder, and sunflowers, the more important crops of maize, beans, and squash did not enter the region until very late, and they did not appear simultaneously. Maize appeared first but did not become a significant food resource in the East until perhaps A.D. 800 to 1000, and beans did not enter the Northeast until about A.D. 1300 (Rieth 2002: 3).

The shift to Late Woodland was characterized by a substantial change in both subsistence and settlement patterns. Sites of this period included villages (some undefended and others palisaded), hamlets, camps (recurrent and temporary), cemeteries (ossuaries), ceremonial dumps, and workshops. Large villages and semi-sedentary occupation appeared, especially in major river valleys, no doubt reflecting growing populations and more predictable modes of subsistence. In New York the Owasco culture practiced the cultivation of maize at the beginning of the Late Woodland, and maize horticulture appeared in western New England at about the same time. Still, other subsistence activities, such as hunting, fishing, and gathering, continued to be important.

Late Woodland sites represented semipermanent habitations, in contrast to the single-component sites identified in the preceding stages of development. The locations of these true habitation sites may have reflected the growing importance of horticulture and the need to select a defendable position. Therefore, sites of this period are located away from the main streams, upon elevated, singular knolls near feeder springs or smaller tributaries.

The Woodland Period in New Hampshire

It is impossible to state exactly when the first palisaded village was erected in New Hampshire, or when the first maize kernel was planted in a farmer's field, or even when the first surplus food was placed inside a storage

FIG. 3.3. Vinette I pottery discovered in feature 5 at the Eddy site in Manchester, radiocarbon-dated to 3,315 +/- 90 years B.P. Courtesy of Victoria Bunker.

FIG. 3.4. A kernel of corn discovered at the Campbell site in Litchfield. The kernel is "charred and hollow in the center," suggesting to Victoria Bunker that this may have been green corn that popped (Bunker 1988: 24). Courtesy of Victoria Bunker.

pit lined with bark for winter consumption. However, we *can* say that Vinette I pottery was being made by a very early date along the Merrimack River, because early specimens have been discovered in radiocarbon-dated features at both the Eddy site in Manchester (fig. 3.3; Kenyon 1986, Bunker 2002) and the Beaver Meadow Brook site in Concord (see figs. 3.1 and 3.2; Howe 1988). These were probably locally made, and no doubt the concept of pottery manufacture spread quickly throughout the state. However, it is doubtful that the lifestyle of Native people in New Hampshire changed appreciably for many years to come. Some models suggest that horticulture may not have existed in New Hampshire prior to European Contact, and farming and palisaded villages may in fact have been a *response* to Contact. If that is so, it would help to explain why no examples of cultigens have ever been found in a prehistoric context in New Hampshire. The only kernel of (possibly) early flint corn found in the Merrimack Valley was discovered at the Campbell site in Litchfield (fig. 3.4), and it may well have come from a Contact period context because a sheet copper disc was found nearby (Bunker 1988: 24–25).

Changing pottery decorating techniques have proven to be a useful basis for dating Woodland period cultures in New Hampshire, and projectile point types are another good indicator. The Early Woodland was marked by Meadowood and Rossville points and by a continuation of Small Stemmed points; the Middle Woodland by Jack's Reef Corner-Notched, Jack's Reef Pentagonal, Greene, and Fox Creek points; and the Late Woodland by the ubiquitous Levanna point (Bunker 1994: 23). After many years of gradual change during the Archaic, the Woodland period in New Hampshire offers a somewhat more dynamic picture, as evidenced by a lively exchange in lithic materials and pottery decorative techniques across the region and by larger settlements with a much richer material culture. The incredibly rich trash pits scattered across the Smyth site at Amoskeag Falls are but one indicator of the growing wealth and sedentariness of Late Woodland times.

The Merrimack Valley

Woodland period ceramics in New Hampshire are best known from sites excavated in the Merrimack River Valley. This is the result of the several large excavations conducted at major falls on the river (Amoskeag Falls, Garvin's Falls, and Sewall's Falls), but it is also due to the pioneering work of Victoria Bunker (previously Kenyon), who performed intensive stylistic and technological attribute analysis on ceramics found at many sites along the river (Kenyon 1981, 1982, 1983, 1985a, 1985b, Bunker 2002). Her work focused on decorative style, aspects of manufacturing technique, and vessel morphology, and it has been through her influence that local pottery studies have come to be increasingly based upon New Hampshire findings rather than ceramic typologies from other states (especially New York and Massachusetts).

The overwhelming quantity of ceramics unearthed at the Smyth site in Manchester in the late 1960s (fig. 3.5) became the basis for much of Bunker's original research. While the lower layers at the Neville and Smyth sites possessed excellent stratification, the upper layers were unfortunately more compressed. The presence of many pits and intrusions made it difficult to isolate components in these shallower layers, and thus the Woodland period is harder to characterize than some of the earlier cultures that resided at Amoskeag Falls.

Bunker was able to identify 767 rims from different pottery vessels at the Smyth site, and she selected rims exclusively for her detailed pottery analysis. All parts of the Woodland period were represented in the ceramic collections, as were Meadowood points from the Early Woodland, Jack's Reef Corner-Notched points from the Middle Woodland, and Levanna points from the Late Woodland.

FIG. 3.5. Howard Sargent digging at the Smyth site in 1968. (Howard is standing at the right rear.) Courtesy of the New Hampshire Archeological Society.

FIG. 3.6. The excavation at the Eddy site. The depth of cultural deposits was about two meters below grade. Courtesy of Victoria Bunker.

Ceramics at the Neville site were much more modest in quantity than at Smyth (a total of about twelve hundred sherds were found at Neville), and Dena Dincauze noted the presence at Neville of Vinette I vessels, Middle Woodland rocker- and dentate-stamped vessels, and "probably Chance-incised" sherds that would have come from the west (Dincauze 1976: 81–84). Basically, the Neville site was no longer all that important during the Woodland period.

Close by, Bunker's work at the Eddy site in 1985 recovered sherds from the Early and Middle Woodland periods, and she obtained a date of 3,315 +/- 90 years B.P. (GX-12385) on Vinette I pottery in feature 5 (figs. 3.6–3.8; Bunker 1992: 138). This exceptionally early pottery at the Eddy site is currently the oldest dated pottery in the state of New Hampshire. And to the south, in Litchfield, Bunker found early Middle Woodland pottery at the Campbell site (figs. 3.9 and 3.10) and late Middle Woodland pottery at the Rodonis Field site (Bunker 1988).

Both of the major falls in Concord have produced sizeable collections of Woodland period ceramics, and excavations at Garvin's Falls at the southern end of Concord have revealed large quantities of Middle Woodland pottery (but only very small amounts of pottery from the Early and Late Woodland) (Kenyon 1985). The SCRAP excavations I directed on the east bank of the Merrimack at Garvin's Falls in 1982

FIG. 3.7. Feature 5 (partially excavated) at the Eddy site. This feature contained the Vinette I pottery that appears in figure 3.3. Courtesy of Victoria Bunker.

FIG. 3.8. A drawing of feature 5 (fig. 3.7). Courtesy of Victoria Bunker.

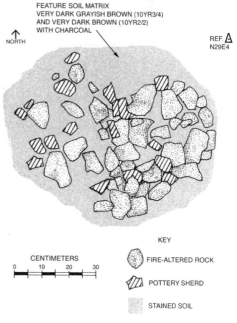

FEATURE SOIL MATRIX
VERY DARK GRAYISH BROWN (10YR3/4)
AND VERY DARK BROWN (10YR2/2)
WITH CHARCOAL

NORTH

REF. N29E4

KEY

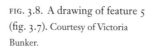
CENTIMETERS
0   10   20   30

FIRE-ALTERED ROCK

POTTERY SHERD

STAINED SOIL

EDDY SITE FIELD DRAWING: FEATURE 5 WITH POTTERY *IN SITU*
EXCAVATION PLAN AT 63-66 CM BELOW SURFACE

FIG. 3.9. The Campbell site in Litchfield. Alan Strauss (left) is excavating pottery, while Pat Hume draws it in. Courtesy of Victoria Bunker.

FIG. 3.10. Looking from the Campbell site toward the Merrimack River. Courtesy of Victoria Bunker.

demonstrated that this site is quite rich in Middle Woodland features. We excavated a large "roasting platform" in area 4 (figs. 3.11 and 3.12; Starbuck 1985) that was radiocarbon-dated to 1,025 +/– 125 years B.P. (GX-9021), so this fell at the very end of the Middle Woodland. Our dig crew posed for repeated pictures while standing behind this massive fireplace (presumably used for smoking fish or otherwise processing food at the falls).

I will never forget, though, just before we finished exposing the rocks, how we were treated to a surprise visit from channel nine television in Manchester. The reporter asked me the usual question: "Have you found anything interesting here?" Almost on cue, one of our SCRAP diggers, Robin Bagley, jumped up, ran around the fireplace, and said, "I think I might have something." The reporter followed me over, and together we examined the complete Levanna projectile point that Robin had just found on the edge of the feature. The reporter asked me whether this was a good find, and I responded "Yes!" so the filming could begin. After all, most artifacts do not "choose" to be found on cue when the television cameras arrive!

We found other Middle Woodland features at Garvin's Falls, including a shallow hearth in area 1 that contained a Jack's Reef Corner-

FIG. 3.11. The "roasting platform" in area 4 at Garvin's Falls. It was overlain by much Middle Woodland pottery, and a Levanna point lay nearby. The platform was radiocarbon-dated to A.D. 925, so it falls in the late Middle Woodland.

FIG. 3.12. A drawing of the "roasting platform" in figure 3.11.

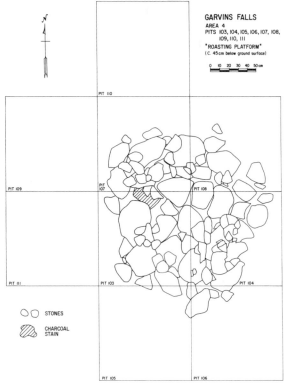

GARVINS FALLS
AREA 4
PITS 103, 104, 105, 106, 107, 108, 109, 110, 111
"ROASTING PLATFORM"
(C. 45 cm below ground surface)

0 10 20 30 40 50 cm

PIT 110

PIT 109   PIT 107   PIT 108

PIT 111   PIT 103   PIT 104

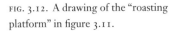 STONES

CHARCOAL STAIN

PIT 105   PIT 106

FIG. 3.13. A shallow hearth in area 1 at Garvin's Falls (Starbuck 1985: 34). This was in association with a Jack's Reef Corner-Notched point and Middle Woodland pottery. There was too little charcoal to obtain a radiocarbon date.

Notched point and Middle Woodland pottery (fig. 3.13). While Garvin's Falls has yielded artifacts from nearly all time periods, it was definitely during the Middle Woodland that it saw the heaviest occupation.

Interestingly enough, the pottery assemblage collected at Sewall's Falls in northern Concord (fig. 3.14) indicates that that part of the river was occupied during both the Early and Late Woodland, as well as the Contact period, but only minimally during the Middle Woodland (fig. 3.15). Of the dozens of hearths we located at Sewall's when we conducted our SCRAP survey there between 1981 and 1984, we submitted only a small number of charcoal samples from hearths for radiocarbon-dating (Starbuck 1984: 7). Two of the hearths dated to the Early Woodland: the first, in sod 8, dated to 1,415 +/- 140 years B.P. (GX-9794; fig.16), while the other, in sod 6, dated to 2,300 +/- 135 years B.P. (GX-9795; fig. 3.17). I believe that many of the undated hearths on the east bank of the Merrimack River pertain to this time period as well.

However, the Late Woodland presence was also substantial at Sewall's, and some very diagnostic pottery sherds were found eroding out of the east bank of the Merrimack River. We repeatedly combed the bank and

FIG. 3.14. The excavation at Sewall's Falls on the Merrimack River.

FIG. 3.15. A sherd of dentate pottery excavated at Sewall's Falls. Drawing by Ellen Pawelczak.

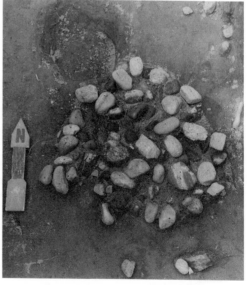

FIG. 3.16. A very shallow hearth in sod 8 at Sewall's Falls (Starbuck 1982: 30). This was radiocarbon-dated to 465 B.C., placing it within the Early Woodland.

FIG. 3.17. A hearth in sod 6 at Sewall's Falls. This was radiocarbon-dated to 350 B.C., placing it within the Early Woodland.

FIG. 3.18. The partially reconstructed rim from an incised pot that we found eroding out of the east bank of the Merrimack River at Sewall's Falls. Drawing by Ellen Pawelczak.

FIG. 3.19. Late Woodland or Contact period rim sherds found eroding out of the east bank of the Merrimack River at Sewall's Falls near sod 9 (Starbuck 1985: 89).

FIG. 3.20. A drawing of a rim sherd in figure 3.19 (upper row, far left). Drawing by Ellen Pawelczak.

FIG. 3.21. A drawing of a rim sherd in figure 3.19 (lower row, far right). Drawing by Ellen Pawelczak.

searched around the tree roots and often were rewarded with decorated rim sherds that were finely made and very late in time, possibly dating to the Contact period, when the sachem Passaconaway sometimes camped with his people at the falls. Some of the better examples of this very late pottery are illustrated in figures 3.18 through 3.23.

While we did not find appreciable samples of Vinette I pottery on the east bank of the Merrimack at Sewall's Falls, we were far more successful on the west bank. We began testing the Beaver Meadow Brook site in 1984, and it was while we were digging shovel test pits that year that we suddenly found rim sherds in the sifting screen. I had been the one digging, and Dennis and Toni Howe were sifting the dirt. I distinctly remember scraping down the sides of the shovel test pit to determine the source of the pottery. It was then that we found the freshly cut edge of a pot in the wall—right where my shovel had gone through it!—and we eventually opened up another five square meters and found some 520 sherds (weighing 1,223 grams) from the same vessel (Starbuck 1984: 100). This was partially reconstructed (see fig. 3.1), and Dennis Howe later drew the possible appearance of the complete vessel (see fig. 3.2), which would have had a diameter of about 26 centimeters and a height of about 28 centimeters. It was definitely a Vinette I vessel, and two other vessels were subsequently found close by; altogether, more than 2.2 kilograms of Vinette I pottery (over seven hundred sherds) were found at the site (Howe 1988: 82). Some charcoal was found in association with the first vessel, and it was radiocarbon-dated to 3,150 +/− 125

FIG. 3.22. Incised Late Woodland pottery sherds excavated from sod 2 at Sewall's Falls.

years B.P. (GX-14011). At the time, it was the earliest pottery ever found in New Hampshire, but later it was surpassed by the date that Bunker obtained on her pottery at the Eddy site. A Meadowood point base was also found at Beaver Meadow Brook, so this site falls very comfortably within the Early Woodland.

FIG. 3.23. A drawing of a sherd in figure 3.22 (left). Drawing by Ellen Pawelczak.

The Lakes Region

Relatively few systematic studies of Woodland period sites have been conducted in the Lakes Region, although many of the multicomponent sites have yielded pottery from both the Middle and Late Woodland. Our 1981 work at NH31-20-5 in Belmont recovered some 34 decorated sherds and 373 undecorated sherds, including a number of dentate and incised pieces; examples of some of these are depicted in figure 3.24. A majority were found in the tightly compacted area of the North Workshop (fig. 3.25), mixed in with Middle Archaic lithic debitage. As is so often the case, the sherds were extremely fragmentary, and no vessels could be reconstructed. Forty-four sherds showed signs of charring, presumably from food preparation (Starbuck 1982: 58–73). One Meadowood biface trianguloid end scraper was recovered from the site, comprising the only evidence for a possible Early Woodland presence.

A short distance away, in Tilton, the Lodge site (NH31-6-6) revealed modest quantities of pottery sherds, and 41 were decorated and 104 were undecorated (Gengras and Bunker 1998: 21–26). Middle Woodland ceramics were most strongly represented at Lodge, although there were small Early Woodland and Late Woodland components as well. These were accompanied by Jack's Reef Pentagonal and Fox Creek points (Middle Woodland) and two Levanna points (Late Woodland).

FIG. 3.24. Pottery sherds recovered at the Belmont site. These show a variety of decorative motifs: *a, b, f, g* are notched; *b* is punctated; *c, e* have thumbnail impressions; *a, b, d, f, g* have horizontal incisions; *a, b, c, d, e, f, g* have oblique incisions; and *d* has concentric incised triangles.

FIG. 3.25. Overview of the excavated North Workshop at the Belmont site. Most of the pottery sherds found at the site were recovered from within this area.

Southeastern New Hampshire and the Seacoast Region

On the seacoast of New Hampshire the most outstanding Woodland period site is the Rocks Road site in Seabrook. The 1973 to 1975 excavation by Charles Bolian and a UNH team recovered very sizeable quantities of ceramics, and these were analyzed by Robert Goodby (Robinson and Bolian 1987; Goodby 1995). Goodby's careful study of the ceramics in the archeology laboratory at UNH identified vessels from the Early, Middle, and Late Woodland and Contact periods.

Robinson and Bolian's original analysis pointed out that many pots were collared, others were castellated, and some have fine incisions and thin body sherds that tend to reinforce a placement very near the end of the Late Woodland or into the Contact period (Robinson and Bolian 1987: 43). Goodby's work was focused more upon ceramic technology, and his detailed attribute analysis identified three vessel lots that belong to the Early Woodland, seventeen that belong to the Middle Woodland, nine that belong to the Late Woodland, and eight that belong to the Contact period. Goodby's work, combined with that of Bunker, marks a significant step forward in ceramic analysis in New Hampshire.

The Connecticut River Valley

The most significant known Woodland period site on the New Hampshire side of the Connecticut River is clearly the Hunter site in Claremont, where Howard Sargent conducted survey work in 1952, followed by excavations in 1967. This was a salvage excavation prompted by a highway relocation and bridge construction, and Sargent was assisted by his students at Nathaniel Hawthorne College and by the Claremont Historical Society.

Working together at the east end of the Ascutney Bridge in Claremont, they discovered some seven occupation levels at this very deep site, with components dating to the Middle and Late Woodland and "evidence of at least three longhouses" (Sargent 1975: 19). Radiocarbon-dating revealed dates of about A.D. 1300 +/- 120 years (GX-1866) and A.D. 1430 +/- 95 years (GX-1854), and Sargent felt that there were strong resemblances between pottery at Hunter and Iroquoian-style pottery, specifically the types known as Oak Hill and Chance (Sargent 1969, 1971, 1975: 19). Thanks to Sargent's efforts, the Hunter site was listed on the National Register of Historic Places in 1976 as "Hunter Archeological Site."

The End of the Woodland

The Woodland period effectively ended when the Europeans brought with them the diseases that killed most Native peoples living in the

Northeast. All the same, it was during the Woodland that recognizable patterns had begun to appear in New Hampshire: the qualities that we associate with the Abenaki of the Contact period were all developing during the Woodland, and when archeologists dig Late Woodland sites in New Hampshire, we are no doubt unearthing the immediate ancestors of the Abenaki.

Our perceptions of Native Americans in the years that followed Contact are greatly altered by the existence of historical sources. Native peoples become infinitely more knowable and "real" as seen through the many first-person accounts of the late sixteenth and seventeenth centuries, and these are the lasting images still with us today.

Chapter 4

The Contact Period

THE NATIVE PEOPLES of New Hampshire were still enjoying the lifestyle of the Woodland period when the first English fishermen and settlers landed on the coast of New Hampshire. Among their more distinctive features, the original residents of New Hampshire practiced horticulture in the summer, relying upon maize, beans, and squash for sustinence. During the winter they dispersed into smaller family units and hunted, and they gathered at falls on the major rivers during the spring for the annual fish runs. They lived in dome-shaped wigwams covered with bark (fig. 4.1), and they moved their villages as necessary when the land or firewood was exhausted. They practiced a matrilocal residence pattern but were patrilineal in how they reckoned descent. But above all, they were an independent, self-governing people whose culture had evolved over thousands of years.

Some primary sources have survived from the Contact period (Calloway 1991), and there is also a vast secondary literature (see, for example, Bragdon 2001, Caduto 2003, Cook 1976, Day 1978, Lottero 1983, and Stewart-Smith 1994), so it is possible to describe the Native people of New Hampshire in the late 1500s and 1600s with some degree of accuracy. There is also a large literature available in the form of town histories, and while these may be uneven in quality, some are quite helpful in describing the lifestyles of Native peoples during the Contact period (see, for example, Belknap 1792, Bouton 1856, Lyford 1903, and Potter 1856).

Abenaki means "people of the dawn," and Western Abenaki was the language of the Native peoples who lived from the White Mountains of New Hampshire west across Vermont to Lake Champlain. It is on the basis of linguistic differences that they are to be distinguished from the more numerous Eastern Abenaki of Maine. The Winnipesaukees (to the north) and the Penacooks were the tribes of the Merrimack River who lived in central and southeastern New Hampshire and into northeastern Massachusetts. During the early Contact period they are sometimes described as being politically separate from the Western

FIG. 4.1. A recreated wig-
wam on the nature trail at
"America's Stonehenge" in
North Salem.

Abenaki to the north—although they were very similar in language and
culture—but after the epidemics that swept New England in the late
1500s and 1600s, they were subsequently identified as having combined
with the rest of the Western Abenaki. The Penacooks later merged with
the Sokoki of the upper Connecticut River and became the St. Francis
(Francois) Indians in Quebec.

It is difficult to reconstruct population figures at the time of European Contact, but the Western Abenaki are usually described as having
had a population of about ten to twelve thousand at their peak, which
dropped to no more than about 250 after they suffered the effects of
disease and war (Snow 1980: 34); this would assume a mortality rate of
98 percent for the Western Abenaki (Snow 1980: table 2.2). Historical
sources indicate that the Penacook traveled among several village sites
and fishing stations along the Merrimack, especially favoring Amoskeag
Falls in Manchester, Sewall's Falls in Concord, and Pawtucket Falls in
Lowell, Massachusetts. Each of these locations has seen the survival of
extensive archeological remains until quite recently.

There is no denying that relations between the Native peoples of
New Hampshire and the European invaders were mixed over time.
Sometimes the Indians of New Hampshire were viewed as "savages"
to be pushed out of the way (they lacked the cities and material goods
of the English), but there were other periods when relations were reasonably amicable for long periods of time (Calloway 1991). Ultimately,
though, those Native peoples who had survived the epidemics moved
north to Canada, and especially to the community of St. Francis (now

known as Odanak), and it was not until the nineteenth century that appreciable numbers of Abenaki began to filter back into New Hampshire. In 1884 Chief Joseph Laurent established a camp in Intervale, New Hampshire, and every summer he led Abenaki people from Odanak to Intervale, where they created handicrafts to sell to the growing tourist trade in the White Mountains. Joseph Laurent's camp in Intervale and the Abenaki Indian Shop operated by his son Stephen were invaluable resources to the people of New Hampshire (Hume 1991; McKenzie 2001). Just as important were the Abenaki dictionary (1995) and other translations prepared over many years by Stephen Laurent, who passed away in 2001.

Today there are about a thousand Native Americans living in the State of New Hampshire, a great many of whom trace their origins to other tribes from other parts of the United States and Canada. The Abenaki do not have Federal recognition as a tribe, which is a source of frustration to many, but they do play a significant role in New Hampshire today, and they do possess a reburial plot in the north country given to them by the State of New Hampshire. Because Abenaki human remains are sometimes discovered by accident, this very private site is where the Abenaki are now able to reinter the remains of their ancestors in perpetuity.

The Indian Chief Passaconaway

The best known of all of the Penacook chiefs or "sachems" in New Hampshire was Passaconaway (fig. 4.2), who had authority over as many as 30 villages, including those at Pawtucket Falls, Amoskeag Falls, and three sites in what is now Concord (Sewall's Falls, Fort Eddy Plain, and Sugar Ball Bluff). Passaconaway is believed to have lived from about 1575 until 1665, and he was buried on Cartagena Island in the Merrimack River. Beginning in 1981, my excavations at Sewall's Falls were intended to locate one of his forts (Starbuck 1982a). My belief was that archeology at a Penacook fort would be an excellent way to document the nature of contact between the two cultures; unfortunately, we have yet to discover any traces of the fort.

Ironically, Passaconaway returned to the news in 1993 when Richard Cogswell of Ossipee drafted a resolution calling upon the U.S. Congress to request the return of Passaconaway's remains from the Museum of Natural History in Paris, where Passaconaway's bones were rumored to have been sent in 1821. Unfortunately, the French could find no record of the skeleton. It would definitely be regrettable if Passaconaway were to rest forever far away from the homeland of his people.

FIG. 4.2. The best-known image of Passaconaway: "Papisseconewa, Sagamon, of Pennacook" (from Potter 1856: 53).

The Archeology of the Contact Period

Only a small number of archeological sites have ever been located in New Hampshire that provide physical evidence for early contact between Native Americans and Europeans. (These sites are summarized in an excellent synthesis by Jane Potter [1998].) Typically these sites have a fascinating mixture of European and Native material culture. Objects of brass and copper, and specifically knives, hoes, tobacco pipes, beads, and firearms, were all highly desired by Native peoples.

One of the best Contact period sites to be excavated in New Hampshire was the Smyth site (NH38-4) at Amoskeag Falls, where a New Hampshire Archeological Society team found "brass and copper arrow points and a cut-out in the form of a bird made of sheet brass [a thunderbird]. Additional items excavated which may possibly relate to Indian use include kaolin pipes, strike-a-lights of European flint, brass scraps, wire, English ceramics, and gun flints" (Winter 1975: 8). A bit further south on the Merrimack River, Victoria Bunker was directing a SCRAP team at the Campbell site (fig. 4.3) in Litchfield in the 1980s when they discovered a perforated metal disc of copper (1.93 by 1.9 centimeters) in association with incised (Late Woodland) pottery (Bunker 1988:15). According to her, "Notably, a Contact period trading post, Cromwell's, was located right across the river and the disc may be contemporaneous" (personal communication, July 6, 2005). Cromwell's had been established by 1656, and so the copper disc (fig. 4.4) appears to have become a Contact period trade item at a very early date.

Another excellent Contact period site on the seacoast of New Hampshire is the Rocks Road site (NH47-21) in Seabrook, where a University of New Hampshire team in 1974 and 1975 revealed "A French gunflint, copper arrow points, a brownware pipe bowl and a deer effigy cut out of sheet lead along with native incised pottery" (Robinson 1983: 2). The Contact period component at Seabrook also included iron axes, two iron knife handles, and the tooth from a bone or ivory trade comb (Robinson and Bolian 1987). Brian Robinson and Charles Bolian believe that these trade items date to between 1600 and 1630.

Two other recently discovered sites provide additional evidence for the Contact period. The Hormell site in Freedom is multicomponent and spans from the Middle Archaic through the early historic period. As noted on its Division of Historical Resources site form (27-CA-15), "It contains an early Contact period occupation represented by European artifacts (brass points, gun flints, kaolin pipe and glass fragments) in context with byproducts of fur processing [fig. 4.5]. This component appears to represent one of the earliest documented Native American historic period occupations in the interior of New England" (Pilkovsky and Boisvert 2004).

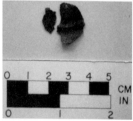

FIG. 4.4. A disc of sheet copper discovered at the Campbell site. Courtesy of Victoria Bunker.

FIG. 4.3. Two SCRAP volunteers at work at the Campbell site in Litchfield in the 1980s. Courtesy of Victoria Bunker.

FIG. 4.5. Three Contact-period brass arrowpoints found in 1993 at the Hormell site (27-CA-15), Freedom, New Hampshire. Courtesy of Richard Boisvert, NHDHR.

The second site is the Conner site (27-CO-34), located on a flood-plain terrace of the Androscoggin River in Shelburne (fig. 4.6). The Conner site was discovered by Victoria Bunker, Inc., during a survey for the Portland Natural Gas Transmission System, and in 1996 Jane Potter found "fourteen modified and unmodified pieces of European flint. Most pieces show battering, polish or worn flake scars. . . . Collectively, they exhibit the variation in form, color, and cortex that is characteristic of European ballast flint" (Potter 1998: 60). The two modified flint pieces appear to have been used as strike-a-lights (one may have been used as a gunflint), and they are fascinating examples of European flint that was first transported as ballast to the New England coast, later to be recycled far inland by Native peoples (figs. 4.7 and 4.8).

Every one of these sites provides important evidence for the process of acculturation as Europeans and Native peoples traded and soon irrevocably altered each other's lifestyles. But none of these discoveries approaches the scope of the research project directed by Peter Thomas at the Sokoki (Squakheag) site of Fort Hill in Hinsdale, New Hampshire. Fort Hill was a Middle Connecticut Valley village occupied from fall to spring of 1663–1664, and it was used by Thomas as the primary case study within his dissertation for the University of Massachusetts, Amherst (1973b, 1979). Thomas excavated some 3,700 square feet of the village site, plus tested a much larger area for features, so his sample size goes well beyond that of any other Contact period village in New Hampshire. His results were equally impressive, as he discovered storage pits and refuse pits, evidence for house layouts and activity areas, ample

FIG. 4.6. A view of the Conner site (27-CO-34), with the field crew taking a break to watch the moose that visited the site daily. Courtesy of Victoria Bunker.

FIG. 4.8. A strike-a-light or gunflint discovered at the Conner site (27-CO-34), modified from European ballast flint. Courtesy of Victoria Bunker.

FIG. 4.7. Unmodified pieces of European ballast flint found at the Conner site (27-CO-34) on the Androscoggin River. This is an important "indicator of Contact-period interaction and travel into the far north of New Hampshire" (Victoria Bunker, personal communication, July 6, 2005). Courtesy of Victoria Bunker.

faunal and floral remains, and a truly impressive variety of European trade goods that had been adopted into the Squakheag village. Many of these trade items would have come, either directly or indirectly, from John Pynchon in Springfield, Massachusetts (Thomas 1979: 363). Some trade items were fairly predictable, such as 39 tobacco pipestems and 27 whole or partial bowls, 1,854 ceramic sherds, some bottle fragments, 200 glass beads (which could have been received from the French), 21 pieces of lead shot, some musket parts, gunflints, and a host of iron objects, including knives, Jew's harps, a hatchet blade, a pair of scissors, a kettle, nails, spikes, and so on. But perhaps the most revealing artifacts of all were seven brass Jesuit rings, very good evidence for contacts with the French in Canada (Thomas 1973a). These rings were probably given to the Indians by Jesuit missionaries as rewards, and they provide solid evidence that the Squakheag of Fort Hill were trading with and becoming increasingly dependent upon *two* cultures other than their own.

Who "Owns" the Past?

Most of the archeological research in the United States is conducted by white Americans of European descent. There are extremely few Native American archeologists or anthropologists, but all of us would like to see many more Native Americans go into our field. In this vein, it would be most helpful for modern-day Abenakis and Penacooks to be part of the process whereby sites are selected and questions are asked about their ancestors. We archeologists welcome Native peoples who participate first-hand on our digs, and we all can benefit by working closely together.

Ironically, a handful of archeologists have advocated that only members of minority communities should be allowed to define "appropriate" research questions or to conduct research on their own ancestors. This restrictive attitude has led to controversy, especially with regard to African American sites, but the vast majority of scholars believes that the past belongs to everyone. We are a pluralistic society, and the very nature of research demands that *all* scholars have equal access to *all* research materials and topics.

Repatriation and NAGPRA

Over the years a small number of Native American skeletons have been unearthed in New Hampshire, sometimes the inevitable consequence of earth-moving activities. In other cases, human remains have been discovered by archeologists who did not necessarily expect to encounter skeletons as they excavated in living sites or middens. Because human remains tend not to last for long periods in New Hampshire soil, most of these remains have dated to the very end of the Woodland period or to the early Post-Contact period, but in all cases they have had the potential to cause highly politicized situations when the descendant communities find that their ancestors have been disturbed.

Issues of disrespect nationally had become so serious that in 1990 Congress passed the Native American Graves Protection and Repatriation Act (NAGPRA), and it was signed into law by President George Bush. NAGPRA requires that all museums and universities receiving federal funds have to provide inventories of Native American sacred objects and human remains to the tribes believed to be culturally affiliated with those remains. Federally recognized Indian tribes can then request the return of those artifacts and remains. Several prehistoric skeletons taken from graves in New Hampshire were affected by this ruling, and Franklin Pierce College and the New Hampshire Archeological Society were among the institutions involved in returning to the Native community the bones that had been found at the Smyth site in 1969.

Just as important is the proper disposition of Native remains that might be found in New Hampshire in the future, and it is a requirement that the state police be notified immediately whenever human remains are discovered. The state medical examiner is then contacted, and ultimately the state archeologist, if appropriate. If the remains prove to be Native American, the state archeologist is required to contact the repatriation coordinator for the Abenaki Nation, and then decisions will be made that determine the final disposition of the bones and any associated artifacts.

One of the most recent examples of this process occurred when human remains were discovered in Holderness (see Goodby 2001). A workman uncovered a human burial on the south side of Davison Brook in April of 2001. Then–State Archaeologist Gary Hume helped to arrange with the repatriation coordinator for preservation of the burials. Then, in June of 2001, additional human remains were discovered in loam that was being spread during improvements to a New Hampshire Department of Fish and Game parking lot. All of the loam (41 truckloads!) was screened by hand, and all bone fragments and artifacts were repatriated to the Abenaki Nation for reburial (Dr. Gary Hume, cited in Goodby 2001: 86–87). Occasional misunderstandings, disagreements, and much expense characterized this entire process—all of which was covered by the media. Nevertheless, respect for human remains has come a very long way, and many individuals from both the Native and the archeological communities worked very hard in Holderness to recover and provide for the reburial of the disturbed human bone fragments.

Part II

The Historical Period in New Hampshire

★ The archeology of the historical period in New Hampshire (post–European Contact) is here divided into three chapters, entitled, appropriately enough, "Historical Archeology," "Industrial Archeology," and "Marine Archeology." There is some potential for overlap—for example, the topic of "blacksmith shops" could be discussed under either the historical or the industrial rubric—but most archeologists who conduct research on historic period sites identify their research as falling within just one of these subfields. As with prehistorians, it is safe to say that historical, industrial, and marine archeologists all share an interest in social behavior, cultural change, and various aspects of technology. But it may be argued that the presence of a solid documentary record enables historic-period archeologists to go much further with their interpretations than prehistoric archeologists can.

Chapter 5

Historical Archeology

What Is Historical Archeology?

ISTORICAL ARCHEOLOGY has been defined as the archeology of European expansion, reflecting the arrival of European culture —and historical records—in such disparate parts of the world as North America and Latin America, South Africa, and Australia. Excavations at historic sites in the United States go back to at least the mid-1800s, as exemplified by James Hall's 1856 excavation of the cellar from Miles Standish's house in Plymouth, Massachusetts (Deetz 1996: 39–40). Hall's record keeping, site map, and plan views qualify his dig as one of the first professional historic site excavations in the United States. However, despite many small cellar-hole digs in the years that followed, historical archeology did not crystallize as a systematic field in the United States until much later, notably when the National Park Service commenced excavations at the early English village of Jamestown, Virginia, in 1934 (Cotter 1958), the Macon Trading Post in Georgia in 1933 (Kelly 1939), the Yorktown Battlefield between 1933 and 1938, and at Fort Raleigh in North Carolina in 1947 (Harrington 1952).

National Park Service archeologists and many others have subsequently dug at hundreds of historic sites, often the homes of famous people or the sites of famous events (such as forts and battlefields), in order to gather the background information necessary to do more accurate site reconstructions and interpretation for the visiting public. Unfortunately, during the early years of this discipline, historical archeology too often had a reputation for *only* studying those whom history had already declared to be important. This image has changed since about 1960, and today the remains of *all* Americans—rich and poor, famous and obscure, European American, Asian American, African American, and Native American—are studied by archeologists.

Much of the early data gathering was rather sloppily done, and much went unpublished, but famous American historic sites provided the training ground for the "first generation" of historical archeologists.

When J. C. Harrington and John Cotter dug at Jamestown, Virginia, and Henry Hornblower II dug at Pilgrim sites in Plymouth, Massachusetts, and Singleton P. Moorehead dug at Williamsburg, Virginia, these pioneers were developing the analytical techniques and research questions necessary to turn this field into a more rigorous discipline (Deetz 1996, South 1994, Derry and Brown 1987). American universities subsequently began to teach college-level courses in historical archeology, beginning at the University of Pennsylvania in 1960 (Cotter 1977). Today, most anthropology departments offer at least one course in historical archeology, and large numbers of historical archeologists are currently employed by universities, state and federal agencies, environmental consulting firms, and historic museum villages such as Colonial Williamsburg, St. Mary's City, and Plimoth Plantation.

While originally much of this work was of a purely descriptive nature, most historical archeologists have succeeded in making the transition from "filling in the gaps" to using archeology as a very real cross-check upon the historical record. Along the way, we have sought to answer questions about human behavior—economic, political, and social—and we have used archeology and material culture studies to identify reoccurring patterns that will demonstrate the values and ideologies of past American culture. Best of all, we have used archeology to give a voice to many of those who were "left out" by traditional history. Historical archeology has given an identity to farmers, laborers, women, and minorities, those groups that have too often been left out of the written record.

Among the earliest historic site excavations in New Hampshire were those conducted by the Civilian Conservation Corps at the Governor John Wentworth plantation in Wolfeboro in 1934 and 1935, as those teams dug inside the cellar hole of the mansion of New Hampshire's last colonial governor (Starbuck 1989). Later, Junius Bird dug a few pits in 1945 at the controversial site of Mystery Hill in North Salem, followed by Gary Vescelius working at that site in 1955; Vescelius conducted a six-week excavation and recovered over seven thousand prehistoric and historic (eighteenth- and nineteenth-century) artifacts. Vescelius was a graduate student at Yale University, and Bird was an archeologist at the American Museum of Natural History in New York City. Both professional archeologists attempted to determine the origin of the stone chambers or "beehives" at the site (Vescelius 1955, 1956). Mystery Hill was then leased in 1957 to Robert Stone, who opened a museum there, and he purchased the site in 1965. Many have conducted excavations at Mystery Hill since then, and the site was renamed "America's Stonehenge" in 1982, reflecting the parallels that the site's owner saw to the great megalithic site of Stonehenge in England and to ancient Celtic

FIG. 5.1. "Contract archeology" often requires tight deadlines, power equipment, and the ability to quickly interpret a wide range of cultural resources, as exemplified by this project in downtown Concord, New Hampshire.

sites in the British Isles and western Europe (see the box "America's Stonehenge").

As these examples suggest, there were few systematic efforts to research historic archeological sites in New Hampshire before the late 1960s, and even national syntheses of this field (for example, Orser and Fagan 1995) have paid scant attention to New Hampshire. This is regrettable, because historical archeology in New Hampshire since the 1960s has been able to research the remains of the first European coastal fisheries, urban settlements such as Strawbery Banke in Portsmouth, the sites of hill farms in the White Mountains, governors' mansions, early factories, potters' shops, and even the site of a large communal society. But over time, excavations prompted solely by research questions have been replaced by those that are necessitated by state and federal laws requiring that sites be identified and recorded before they are destroyed by modern construction. "Contract archeology" now accounts for most of the historic site excavations being conducted in New Hampshire (fig. 5.1), and final reports describing this work are filed with the New Hampshire Division of Historical Resources in Concord.

Excellent research opportunities have existed because abundant physical remains have survived from New Hampshire's historic past, and because many institutions and agencies—especially the Division of Historical Resources—have helped to support this research. However, most historic digs are modest in scope, and at a time when archeological resources are being threatened throughout the United States, it is essential to strike an appropriate balance between the needs of historic

★ America's Stonehenge

Easily the best-known archeological site in New Hampshire is "America's Stonehenge," formerly identified as "Mystery Hill Caves." Located just off Route 111 in Salem, this assemblage of about 22 manmade stone chambers ("beehives"), standing stones, walls, and drains is unquestionably provocative, puzzling, and, above all, controversial. A host of researchers, beginning in 1936 with William B. Goodwin, have tried to identify the builders of this site, to interpret its meaning(s), and to physically document the ruins and stones that are scattered across many acres (see Hencken 1939, Goodwin 1946, Stone 1971, Feldman 1976, and Fell 1989).

A visit to America's Stonehenge is well worth the admission fee, even for the most sceptical, because it cannot be fairly evaluated without seeing all of the evidence up close. There are some good "mysteries" here—and the site *is* thought provoking! From the modern visitor's center it is possible to walk the nature trail, hike up onto the high ground where the stone chambers and observation tower are located, and then walk the perimeter of the site where standing stones (from five to nine feet tall) suggest possible astro-

The entrance of a stone chamber at America's Stonehenge.

Another stone chamber at America's Stonehenge.

nomical alignments. The featured attractions are the so-called "oracle chamber" and the flat, grooved granite slab called the "sacrificial table," both located within the central cluster of stone chambers that covers about one acre. There is also an "astronomical viewing platform" from which sightings may be made to standing stones that are aligned with true north and with the summer and winter solstice sunrise and sunset.

But what exactly does all of this mean? Interpretations differ greatly. William Goodwin purchased 20 acres of the site in 1935 and subsequently rearranged many of the stones. It is widely believed that Goodwin may have "created" much of what is visible at the site

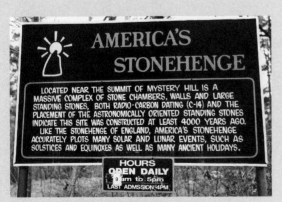

Sign greeting visitors at the entrance of America's Stonehenge.

The "sacrificial table" at America's Stonehenge, weighing 4.5 tons.

An astronomical alignment stone at America's Stonehenge.

today, but nevertheless he believed that Irish Culdee Monks in the tenth century A.D. had constructed the "caves" here as a monastery. However, most professional archeologists believe that a ninteenth-century owner and farmer (Jonathan Pattee) and his family probably erected the stones and constructed the chambers. An alternative explanation is provided in the literature given to site visitors, where the owners of the site claim that "America's Stonehenge is most likely the oldest man-made construction in the United States (over 4,000 years old)." Because I am an archeologist, I'll admit that I tend to believe that Pattee and Goodwin built everything, but that is chiefly because the easiest, simplest interpretation is usually the correct one. Any interpretation that requires ancient diffusion from Europe, Culdee monks, writers from the Iberian Peninsula, or a 4,000-year-old, wandering, megalithic Bronze Age culture simply requires a huge "burden of proof" that has yet to be forthcoming. The nineteenth century was a time when theories of diffusion were really quite popular, but scholars of the more sceptical twentieth and twenty-first centuries tend not to take this sort of thing very seriously—not without tangible evidence of artifacts or raw materials that could only have come from Europe (see Feder 2006).

The biggest problem for many of us is that America's Stonehenge has been so altered by past stone quarrying and by too many well-intentioned researchers attempting to "put the stones back" and "restore" the site to its original appearance. The moment the first stone was moved to a new location by William Goodwin, the entire site lost any chance of being taken seriously by the scholarly community. We academics have sometimes been accused of not being open-minded when it comes to claims about ancient megalithic cultures, and of not respecting "alternative" views of the past. Yet site integrity is everything to an archeologist, and this site is severely compromised.

I am reminded of the first time I visited the "caves" back in the late 1970s, when the name was still Mystery Hill. My guide told me that the whole thing—stone chambers, trails, and so on—was actually the representation of a gigantic Indian or Asian face wearing a peaked hat; that this was an ancient mental concept that had crossed the Bering Straits ten thousand years ago; and that this image had been repeated in Indian art all across America for thousands of years. And now this ancient idea had been displayed, in gigantic size, at Mystery Hill. The next time I visited the site was in 1982, and I was shown ordinary fieldstones that a new generation of diggers was identifying as "manos," "metates," and so on—every last stone on the hillside was being imagined into an artifact! Every time another absurd theory is added to the mix, it becomes harder to accept *any* of the elaborate tales told about this site.

Another significant problem is the dating of America's Stonehenge. Several radiocarbon dates have been obtained on miscellaneous pieces of charcoal and used to proclaim that this site is pre-colonial, even ancient. The dates are real, but the samples do not appear to have very good associations with anything of human origin. And then there are the quarrying marks (drill marks) on the stones, which are clearly of post-1830s origin; the various straight-line scratches and grooves on other stones that have been identified as bearing the ancient Ogam script (from the Iberian Peninsula); and the conspiracy theories that claim Gary Vescelius actually discovered "ancient" artifacts in 1955 and deliberately lost them after he went to the American Museum of Natural History.

I once saw it claimed in writing that possibly the megaliths of Europe are *derived* from America's Stonehenge and sites like it in the New World. All of this is a bit too much to take. If an early site truly has merit, it doesn't require bizarre interpretations. All the same, America's Stonehenge raises interesting questions about early astronomy and stone-working, and it provides a nice cautionary tale about what is or isn't acceptable in modern archeology. Perhaps its greatest value is to challenge us, to force us to demand very high standards of proof before we accept strange theories about the past.

FIG. 5.2. Sherburne Wharf, the reconstruction of an eighteenth-century wharf at Strawbery Banke.

preservation and pure research. Historic preservation need not mean leaving everything in the ground, where it will continue to deteriorate and become harder to interpret. At the same time, pure research should not be used as an excuse for completely digging and thereby destroying sites that have otherwise been undisturbed until now.

Urban Historical Archeology in the City of Portsmouth

Portsmouth was originally settled in the 1630s and was the capital of New Hampshire during the colonial period. Strawbery Banke, which is the oldest part of the city, is considered to be one of the best urban archeological sites in America. Today the Strawbery Banke Museum, founded in 1964, exhibits 42 historic buildings on ten acres of restored gardens and yards, as well as hosting several archeological sites (fig. 5.2). The first archeology did not begin there until 1964, and during that year several members of the New Hampshire Archeological Society excavated in the area known as "Puddle Dock." This was the first of many efforts to document urban development in that city. The Pitt Tavern in Portsmouth also saw the excavation of a privy by a student, Jan Herman, in 1964 or 1965. In 1966, the "pick and shovel archeologist" Roland Robbins continued excavations at Puddle Dock and found large timbers from an early wharf as well as many nineteenth-century bottles. Soon after, in 1968 and 1969, Daniel Ingersoll, then a graduate student at Harvard University, conducted an extensive excavation at Puddle Dock and recovered mainly nineteenth- and twentieth-century fill; Ingersoll also exposed a wharf structure built between 1830 and 1840 (Ingersoll 1971a, 1971b). Portsmouth has seen archeological digs nearly every year since then.

From 1975 through 1978, Steven Pendery excavated the Marshall Pottery site at Strawbery Banke (Pendery 1978, 1984, 1985, DeCorse 1978–1979), and the outline of that site is still clearly visible on the surface of the ground today (fig. 5.3). Pendery also excavated in the Market Square area in downtown Portsmouth in 1976, and in the yards around several historic houses, including the Peacock House in 1975, the Cotton House from 1976 to 1978, the Sherburne House in 1977, the Jefferson House in 1978, the Jones House from 1977 to 1978, and the Yeaton-Winn House in 1978.

Pendery's excavations at the Marshall Pottery site (in use between 1737 and 1749) were easily the most significant, because this was the first professional excavation of any pottery site in New England. Prior to Pendery, collectors had merely dug up sherds from different shops to get a sense of their product lines (Watkins 1950). Samuel Marshall,

FIG. 5.3. A signboard and stones mark the foundation of the Marshall Pottery site at Strawbery Banke.

the potter who had built his house and shop on the southwest corner of Horse Lane and Jefferson Street, used local clay to produce a variety of ceramic containers for Portsmouth and the larger region. The dumps associated with pottery sites typically contain many "wasters," poorly fired pottery pieces that have to be discarded. Pendery's "feature 20," found during his excavation of Marshall's Pottery, was a waster pit associated with a possible kiln. In addition to great quantities of waster sherds, kiln bricks, and decorated tablewares, Pendery found an earth surface with impressions that suggested where wooden racks may have been set to support redware while it dried (Pendery 1981).

Pendery also began digging in 1980 at the Dodge Pottery site (in use between 1789 and 1864) in the North End of Portsmouth. Winthrop Bennett had begun a redware pottery shop there in 1789, and then Joseph Dodge of Exeter purchased the shop in 1796. Dodge and his two sons operated the business until 1864, after which pottery was no longer produced in Portsmouth (Pendery 1985). Pendery's excavation at the Dodge Pottery site was on a much smaller scale than at the Marshall Pottery site, but taken together these sites have provided a wealth of information about eighteenth- and nineteenth-century pottery production in Portsmouth.

In 1979 archeology in Portsmouth was placed on a more permanent footing when the Jones House became the Strawbery Banke Archaeology Center and began to house artifact collections from throughout the city (fig. 5.4). Since that time, a public laboratory and exhibits have helped to share archeological results with the visiting public, even as staff members from the center have conducted excavations in

Portsmouth. The 1980s saw a very considerable increase in archeological activity throughout the city of Portsmouth as Faith Harrington dug at the Follett site (Puddle Dock) from 1981 to 1982 (Harrington 1983a, 1983b) and continued work in the yards around the Sherburne House in 1983–1984 (Harrington 1989; fig. 5.5). Gray Graffam dug at the Rider-Wood House in 1981, using over 120 volunteers, and that year also saw the beginning of the massive Deer Street Project in Portsmouth's North End, some distance from the Strawbery Banke Museum.

Urban renewal had removed many buildings from the North End between 1969 and 1971. As a prelude to new construction, archeological

FIG. 5.4. Jones House, home to the Archaeology Department at Strawbery Banke.

FIG. 5.5. The seventeenth-century Sherburne House at Strawbery Banke, a site of frequent excavations.

work began there in 1981 under Steven Pendery, and he was succeeded later that year by Aileen Agnew, who continued excavations through 1986. A general survey of 10 acres was followed by Data Recovery Phase III excavations at the Hart-Shortridge house lot in 1981 and 1982 (Edwards, Pendery, and Agnew 1988), along with digs at the Richard Shortridge, Richard Hart, and Deer Tavern sites. Agnew discovered extremely rich privy features, stone-lined cellars, wells, trash pits, a brick cistern, and sheet refuse, all of these containing large quantities of glass and ceramics. The Deer Street Archaeological Projects became the largest and most successful urban archeology ever conducted in New Hampshire (Agnew 1985, 1988, 1989, 1995a, 1995b, Pinello 1989, Wheeler 1999, 2000). The analysis of the thousands of artifacts from Deer Street continued well into the 1990s.

From the mid-1980s to the present, excavations at urban sites in Portsmouth have continued unabated. In 1986 the amount of activity prompted Strawbery Banke Museum to hire Martha Pinello as museum archeologist and Mary Dupre as assistant archeologist. Over many years, both archeologists have successfully maintained a very active program of field work and analysis (see Pinello 1989, Dupre 1995). Their work at the Rider-Wood House in 1989 and 1990, at the Wheelwright House in 1990 and 1991, at the Wentworth-Coolidge Mansion in 1992, at the Warner House between 1993 and 1998, and at other sites since then has been part of a concerted effort by Strawbery Banke Museum to keep up with the rapid pace of construction and the need for interpretation and public education in Portsmouth. Pinello's most recent work has been at the Shapiro House, where she has interpreted Jewish immigrant family life.

Separate from the Strawbery Banke Museum, the firm Independent Archaeological Consulting (IAC) has conducted many excellent projects in Portsmouth over the past 15 years. IAC is headed by Kathleen Wheeler, working with her colleague Ellen Marlatt. Kathy Wheeler completed her dissertation on research at the Rider-Wood, Wheelwright, and Follett sites (1992). In the years that have followed, Wheeler and Marlatt have undertaken many excavations in the city, including the 2003 recovery of eight sets of human remains from what was formerly the eighteenth-century "Negro Burying Ground" in Portsmouth (about 1705–1790s). They removed hexagonal wood coffins and their human contents from this site, located at the intersection of Court and Chestnut Streets, and these remains have undergone intensive analysis since then.

Recent work in Portsmouth has also included a study by Carolyn White of eighteenth-century artifacts of personal adornment (clothing fasteners, buttons, buckles, and so on) that were recovered during excavations at the Warner House (White 2002, 2004, 2005). Other work includes excavations in the summer of 2004 that were conducted by a Strawbery Banke Museum team at the Marshall Pottery site, the first work since Steven Pendery's excavations of the 1970s. This recent effort made it possible for a whole new generation of interns and volunteers to work with the redware, wasters, and kiln furniture from this pottery site.

Archeological projects in the city of Portsmouth have now been conducted for more than 40 years, under many different directors, and clearly Portsmouth is the most intensively researched urban area in New Hampshire. Much of this research has been of exemplary quality, and Strawbery Banke Museum in particular is to be congratulated for making such a substantive contribution to urban archeology.

Rural and Farm Sites

While urban archeology was expanding in the 1980s in New Hampshire, work at rural sites was too, beginning with a study of farmsteads conducted by John Wilson in southwestern New Hampshire in 1979. In a cultural resource survey of Surry that involved no excavations, Wilson tackled the central archeological question of significance, using his survey of farmsteads as a springboard for asking "what makes farms significant?" (Wilson 1990). With hundreds of abandoned farmsteads all across the state of New Hampshire, how do archeologists and preservationists decide which ones are typical, which ones are unusual, and which ones have the potential to reveal meaningful new information if they were dug?

Wilson's study was followed by excavations that I conducted in 1980 at the Jones Farm (the New Hampshire Farm Museum) in Milton (fig. 5.6). Soon afterward, Richard Waldbauer, then at Brown University, initiated the "Hill Farm Project" in order to study clusters of hill farms in the White Mountain National Forest of New Hampshire and Maine (Waldbauer 1985, 1986). Waldbauer's project involved a small amount of digging in 1983 at ninteenth-century hill farms (through the auspices of Plymouth State College and the United States Forest Service), but it was chiefly a historical and spatial analysis of "regional agricultural economy of nineteenth-century farms located at the highest elevations in the Northeast" (Waldbauer 1985: 67–68). Waldbauer interpreted a sample of 130 farm sites, noting that hill farms must be studied as *clusters* of farms that were interdependent within a regional economy.

Also in 1983, a project with both rural and industrial aspects was conducted by Faith Harrington, Fred Bissen, and a Keene State College team in Pisgah State Park in Winchester. This had been the site of a lumbering community (the "Broad Brook Steam Lumber Mills")

FIG. 5.6. A 1980 excavation at the Jones House in Milton.

from the 1830s to the 1930s, and dams, ponds, ditches, and cellar holes of dwellings had survived. Harrington dug at the site of a sawmill in the vicinity of a blacksmith shop and next to a mid-nineteenth-century cellar hole (Harrington 1984).

In 1988 Karl Roenke became the archeologist for the White Mountain National Forest (WMNF), and since that time he has directed a systematic program of rural excavations throughout that region of New Hampshire. The WMNF is composed of over 720,000 acres in central New Hampshire and over 49,000 acres in western Maine. To date, Karl and his associates have completed over 250 Cultural Resource Reconnaissance Reports and have located over 1,100 cultural sites on forest-managed lands (fig. 5.7). In addition to required cultural surveys and reports in support of Forest Service projects—such as timber sales, recreation development, road and trail maintenance/construction, and watershed improvement projects—the WMNF organization has offered field schools in historical archeology at several locations, including the historic 1830s Russell-Colbath House on the Kancamagus Highway in Albany, New Hampshire. These field schools were accomplished through partnerships with Plymouth State University and the New Hampshire Division of Historical Resources. In 2004 the forest undertook its fifth consecutive historical archeology field school with the Girl Scouts of the United States of America, Swift Water Council, offering a week of archeological instruction at selected forest sites (fig. 5.8).

FIG. 5.7. Much of the land that became the White Mountain National Forest was intensively impacted historically for its natural resources. The forest contains minerals that were mined during the ninteenth century. This shaft is located at a mine in Warren, New Hampshire, where zinc, copper, and silver were mined from about 1834 to 1915. Courtesy of the White Mountain National Forest.

FIG. 5.8. The White Mountain National Forest is the steward of three historic nineteenth-century farmhouses that are listed, or have been determined eligible for listing, on the National Register of Historic Places. These structures, their surrounding landscapes, and their associated archeological deposits are valuable resources for understanding New Hampshire's agricultural history. The Smith House (built about 1860), also known as the Mead Base Conservation Center, in Sandwich, New Hampshire, was the site of the 2005 WMNF and Girl Scouts of the USA field school in historical archeology. Courtesy of the White Mountain National Forest.

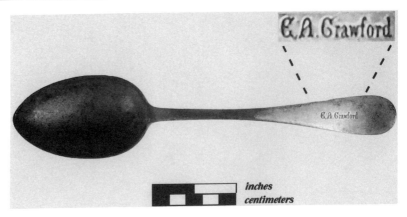

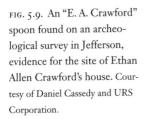

FIG. 5.9. An "E. A. Crawford" spoon found on an archeological survey in Jefferson, evidence for the site of Ethan Allen Crawford's house. Courtesy of Daniel Cassedy and URS Corporation.

While long-term regional survey efforts are always preferable, it is nevertheless common for rural sites to be encountered during cultural resource management projects. One recent example was on Route 2 in Jefferson during a New Hampshire Department of Transportation survey. The Crawford family was one of the very first to settle in the western part of the White Mountains, beginning with Abel Crawford in the late 1700s. Abel's sons subsequently operated inns and escorted visitors to the top of Mount Washington. It was in about 1870 that Abel's grandson, Ethan Allen Crawford II, moved to Jefferson and established a boarding house that was later taken over by his sons, Ethan Allen III and Fred. One hundred and thirty years later, archeologists conducting a survey found many artifacts there from the occupation of the Crawford House/Hotel, but the most noteworthy artifact was a silver-plated copper alloy tablespoon with the inscription "E. A. Crawford" engraved on the obverse handle (Daniel Cassedy, personal communication, June 17, 2005; Cassedy and Parson 2003). It is all too rare that we find "personalized" artifacts, but it certainly is exciting to recover a spoon inscribed with the name of one of New Hampshire's most illustrious families (fig. 5.9)!

An Early Workers' Village

One of the most extensive excavations ever conducted in New Hampshire took place upon the side of Kidder Mountain in the town of Temple between 1975 and 1978. I will never forget my first visit to the site of the New England Glassworks (1780–1782) in the summer of 1975 (fig. 5.10). I had just received my doctorate from Yale University, and Boston University had hired me only a few weeks earlier to co-direct a field school at the site of this early glass factory. I found myself hiking up a

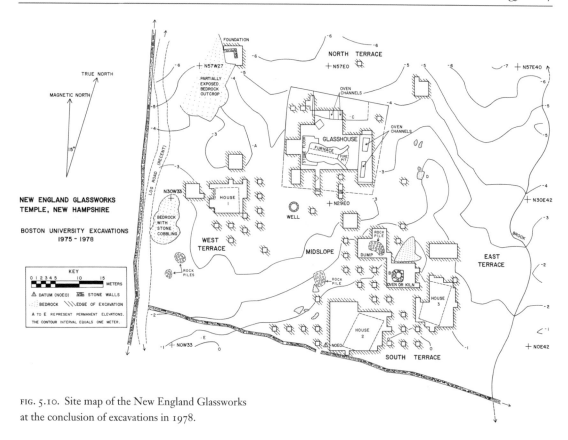

FIG. 5.10. Site map of the New England Glassworks
at the conclusion of excavations in 1978.

log road, looking for the factory ruins and for a chance to meet one of
my new co-directors, Frederick Gorman. As I approached a freshly cut
clearing in the forest, I could smell insect repellent wafting through the
trees, coming from the direction of Fred and about 20 students, who
were sweaty, mosquito bitten, and extremely frustrated because they had
been chopping down trees and brush for days instead of doing the digging they had been promised. Consequently, they had all gone on strike
just minutes before I arrived! Things did get better, but I remember well
the 29 straight days it rained that fall (most of us had walking pneumo-
nia), the student who got poison ivy when she went to the bathroom in
the woods, the "rumble" at the roadhouse in Peterborough (where our
students slugged it out with some of the "locals"), and the student who
went home every weekend to bring back drugs for some of the others on
the crew. Archeology has got to be easier than this!

The Boston University team's excavation of the New England Glassworks site was one of the few times that a small eighteenth-century factory village has been excavated anywhere in America. We exposed much
of the glasshouse (discussed in the next chapter), but we also uncovered

FIG. 5.11. The foundation of house 1 at the New England Glassworks.

the remains of three large workers' cabins (fig. 5.11), a lean-to, and various dumps. All that survived of the cabins were low scatters of field stones and clay, located a short distance south and southwest of the glasshouse (Starbuck 1976–77, 1978, 1983a, 1984a, 1986a). We even brought in ground-penetrating radar in an effort to find additional, ephemeral structures (fig. 5.12).

The workers' village was reputed to have been occupied by "Hessian deserters from the British army," escapees from the Battle of Saratoga in 1777, so we were especially eager to find evidence for the origins of the workers. We thus were quite delighted when we found three British Sixty-third Regiment buttons in the foundation of our "house 1" (fig. 5.13); naturally the buttons generated endless speculation about how they might have gotten there! We never found the gold that the "Hessian" workers supposedly buried underneath one of the fireplaces, but we did find many artifacts inside the cabins, including sherds of redware, creamware (fig. 5.14), slip-decorated earthenware, porcelain, and delft, as well as knives, buttons, buckles (fig. 5.15), cast iron pot and kettle fragments (fig. 5.16), and ox shoes, suggesting a life that must have been relatively spartan, far from home and loved ones. We know that the factory workers relied upon cows, pigs, sheep, and fish for part of

FIG. 5.12. Ground-penetrating radar in use at the New England Glassworks in 1976. Geophysical Survey Systems, Inc., of Burlington, Massachusetts, carried out the survey in an effort to find additional house foundations.

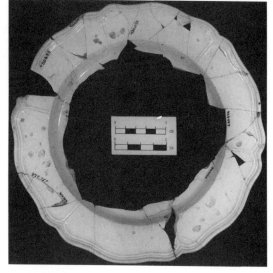

FIG. 5.13. A British Sixty-third Regiment of Foot button (left) and a "USA" Continental Army button, both found at the New England Glassworks. Drawing by Ellen Pawelczak.

FIG. 5.14. A reconstructed creamware plate, Queen's pattern, from house 2 at the New England Glassworks.

FIG. 5.15. Three buckle fragments found in house sites at the New England Glassworks. Drawing by Ellen Pawelczak.

FIG. 5.16. Cast-iron pot and kettle fragments found at the New England Glassworks.

their diet, because we found bones from those animals in dumps behind their houses.

The factory failed after just two years, and the workers apparently returned to Boston, where the factory owner, Robert Hewes, owned a slaughterhouse and tannery. Hewes never repaid a loan of 3,000 pounds received from the selectmen of Temple, and the first time we reported our archeological findings to a meeting of the Temple Historical Society, Fred commented (half seriously) that the townspeople looked like they expected us to repay the 195-year-old loan!

Archeology at a Communal Society

The 1970s also saw the beginning of a massive survey effort to document the history, architecture, and archeological remains of Canterbury Shaker Village, the site of New Hampshire's largest and most long-lasting communal society (from 1792 to 1992). I have directed this effort since 1978, and this lengthy project has given me and my colleagues the opportunity to produce an extensive archive of measured drawings, photographs, and site reports that represents the most thorough body of documentation for any communal society in the United States (Starbuck 1990a, 2004).

The initial field work in Canterbury, using paid students each summer, lasted just three seasons (1978–1980), but the drafting of maps and preparation of written reports continued for much of the 1980s. During the early years, most of our work in Canterbury consisted of surface recording and archival research. We mapped fields, ponds, orchards, stone walls, cellar holes, and buildings (fig. 5.17). The only digging during that period was in crawl spaces underneath buildings (fig. 5.18); in selected features within the manmade mill system (see the next chapter); and at the West Family site, where some of the cellar holes were on the verge of being destroyed by modern farming. There was little immediate need for excavation at the beginning of our project, because the "big picture" could be more easily determined through surface mapping and historical records (see Starbuck 1980a, 1984c, 1986b, 1988a, 1990a, 1990b).

It was not until 1994 and thereafter that research priorities shifted so as to include digging into the Shaker dumps, revealing that the Shakers were great consumers of nearly all that "the world's people" had to offer. In 1994 my students and I dug into the trash-filled cellar hole of a garden barn at the Shaker Church Family, followed by digging in the remains of a Church Family hog house in 1996 and 1997 (fig. 5.19), in the ramp leading into the Church Family cow barn in 1998, and in two blacksmith shops (at the Second Family and the Church Family) between 1996 and 2002. Patent medicines, liquid food, attractively decorated ceramics (figs. 5.20 and 5.21), thousands of tobacco pipes (fig. 5.22), bottles for beer, wine, and whiskey (figs. 5.23 and 5.24), and perfume bottles (fig. 5.25) were among the many findings. Yet our work helped to establish how very much like us the Shakers were—for us they were not romanticized icons and representatives of a "simpler" way of life but real people who bought and consumed and enjoyed life in many of the same ways as the rest of us do (Starbuck 1997, 1998, 1999b, 2000b, 2004).

The Canterbury project has gradually become one of the very best examples of why archeology is able to go beyond traditional history, revealing some of the ways in which people differ from the rosy stereotypes told about them. So much popular mythology has grown up around the Shakers that it is often hard to differentiate between the stories the Shakers created about themselves (and the myths told in "popular"

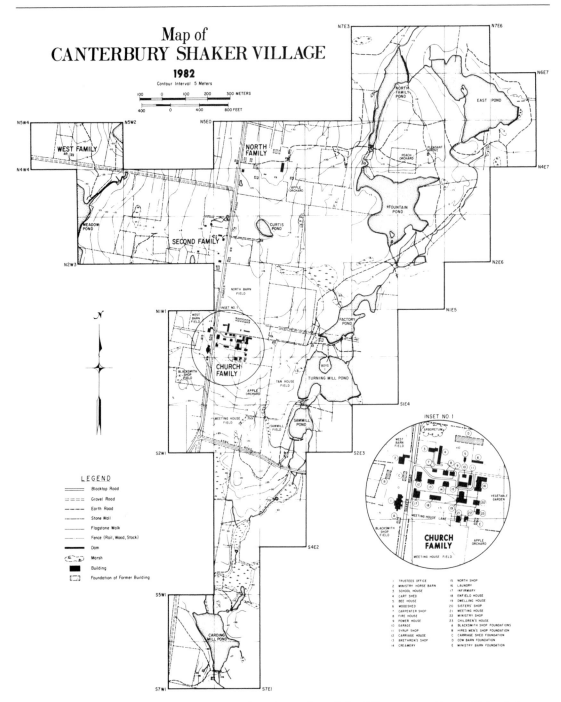

FIG. 5.17. The final map showing the modern-day surface of Canterbury Shaker Village, prepared by combining 61 base maps, each measuring 200 by 200 meters.

FIG. 5.18. The archeology team digging inside the crawl space underneath the Shaker schoolhouse in 1980.

FIG. 5.19. The hog house site at Canterbury Shaker Village near the completion of excavation in 1996.

FIG. 5.20. A yellowware nappy (baking dish) with Rockingham decoration found in Hog Heaven in 1996; 12¼ inches in diameter.

FIG. 5.21. The base of a temperance plate showing Father Matthew "ADMINISTERING THE TOTAL ABSTINENCE PLEDGE" to his followers; this was discovered in Hog Heaven in 1996.

FIG. 5.22. Redware pipe bowl fragments found behind the Second Family blacksmith shop at Canterbury Shaker Village.

FIG. 5.23. A whiskey bottle and a beer bottle found in the garden barn foundation at Canterbury Shaker Village in 1994.

books) and the realities as revealed through their journals and material culture. The many years of archeology conducted at Canterbury Shaker Village have stripped away some of that sanitized image and revealed a somewhat conventional agricultural and industrial community that worked very hard and mirrored nearly all of the changes going on in the outside world. The contents of Shaker dumps are a truer reflection of Shaker customs and changing Shaker lifestyles over two centuries than what is contained in most of the coffee-table books written about them. In the dumps we have found fine ivory toothbrushes, a bottle of "Mrs. A. Allen's World's Hair Restorer," ivory napkin rings, bottles of Saratoga Congress Water, combs, hooks from foundation garments, and lots of "Shaker Cherry Pectoral Syrup" bottles. I for one would much rather study this material evidence than read the long lists of inventions that are sometimes credited to the Shakers by their avid supporters!

Military Sites Archeology

In many states, forts and battlefields are among the most popular, most visited, and most overdug sites of all, reflecting the public's fascination with military history and the events that helped to form our country. While no major battles took place in New Hampshire, several small forts were built along the Merrimack and Connecticut Rivers in the eighteenth century, and several more permanent fortifications were

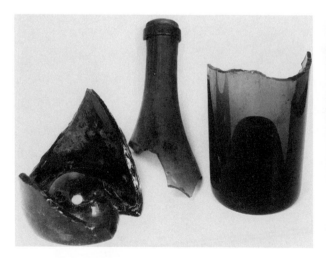

FIG. 5.24. Some of the wine bottle fragments discovered inside the east ramp of the Church Family cow barn at Canterbury Shaker Village.

FIG. 5.25. Perfume bottles discovered inside the east ramp of the Church Family cow barn at Canterbury Shaker Village.

built in the coastal area from the 1600s on, including Fort William and Mary, Fort Constitution, and Fort Stark in New Castle, and Fort Washington in Portsmouth. Howard Sargent did small-scale digging at Fort Constitution in 1968 and 1969, and Faith Harrington did some core sampling at Fort Stark in 1983. A number of earthworks have survived at these military installations, but most of the remains are from the twentieth century. There are only very fragmentary remains from earlier fortifications.

Perhaps the most well-known fort in New Hampshire was Fort No. 4 in Charlestown on the Connecticut River, which was erected to provide English settlers with protection from the French and Indians (it was active from about 1743 to 1763). Howard Sargent dug at what he believed to be the palisaded fort site in 1957 and 1958, where he found postholes and a trench that he hoped might be from the original fort. His findings helped lead to the reconstruction of Fort No. 4 soon afterward. Quite a few years later, Paula Zitzler and a Keene State College team also dug at the possible site of the fort in 1984. Under the general guidance of Faith Harrington, Zitzler opened several pits just north of where Howard Sargent had earlier dug on Main Street and found several eighteenth-century artifacts, including a 1733 British halfpenny, a musket ball, a shoe buckle, and pottery sherds (Harrington 1985a).

Beginning in 1982, a much more definitive fort excavation was conducted in central New Hampshire in the town of Boscawen, close to

FIG. 5.26. Granite boulder erected by the Town of Boscawen, indicating the "Site of First Fort / A.D. 1739 / One Hundred Feet Square / Built of Hewn Logs."

FIG. 5.27. Overview of the excavation area at the First Fort in Boscawen, showing the rocks and clay exposed during Mary Dupre's excavation between 1983 and 1985.

where the Contoocook River flows into the Merrimack River. There were early forts in both Boscawen and Canterbury, and historical records indicate that the proprietors of Boscawen voted in 1739 to build a log fort that would be 100 feet square (fig. 5.26). Still, historical records were vague as to whether the fort was ever actually built. This prompted a preliminary survey by me in 1982, followed by a sizeable excavation of the "First Fort" that was directed by Mary Dupre in the summers of 1983, 1984, and 1985. While no traces of a palisade wall were discovered on this sandy terrace above the Merrimack River, a dense scatter of rocks, bricks, and clay (fig. 5.27)—accompanied by many mid-eighteenth-century artifacts (fig. 5.28)—revealed that these were the remains of a building from the desired time period (Dupre 1985a). Among Dupre's more distinctive findings was the complete skeleton of a dog, undated but possibly buried as early as the time of the fort. The dog was discovered in 1985 within the dense scatter of rubble (fig. 5.29).

Because the exact layout of the fort could not be determined, I resumed work at the "First Fort" in 2003 with the sponsorship of Plymouth State University (PSU) and assisted by PSU student Stephanie Tice. More parts of the fort site have now been tested archeologically, and Tice has recovered many more mid-eighteenth-century artifacts. While there is still no evidence for the fort's outline, Tice has discovered

FIG. 5.28. Some of the artifacts recovered by Mary Dupre at the First Fort, including a bone comb (top left), a leather heel (top center), a mouth harp (top right), a gunflint (center left), a musket ball (center right), pewter buttons (bottom left), and two unidentified metal loops (bottom right).

FIG. 5.29. A dog skeleton excavated at the First Fort in 1985.

more clusters of rocks, probably from fireplaces. It may well require stripping the entire site down to the top of the subsoil before a pattern of palisade posts, ditches, or foundations is finally located, but the "First Fort" in Boscawen unquestionably has the most integrity of any known eighteenth-century fort site in New Hampshire.

A Redware Pottery Shop and Kilns

Excavations have been conducted at the sites of several redware potters in New Hampshire, notably those directed by Steven Pendery in Portsmouth, but the largest dig of all was at the site of a potter in the Millville District of Concord (land now owned by St. Paul's School). Ample supplies of clay suitable for making pottery had long been procured from clay beds in this area, and the third potter to manage one of the shops in this district was Joseph Hazeltine, who occupied a pot shop and built kilns just north of Hopkinton Road (fig. 5.30). Hazeltine—who was in operation from sometime before 1842 until 1880—used the four wheels in his shop to produce a variety of wares, including lard pots, bean pots,

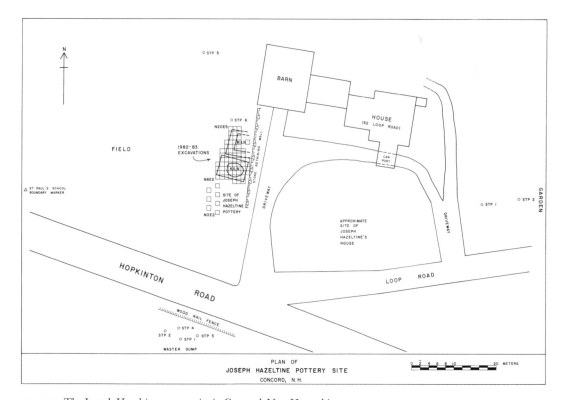

FIG. 5.30. The Joseph Hazeltine pottery site in Concord, New Hampshire.

FIG. 5.31. The excavation of kiln house I at the Joseph Hazeltine pottery site. Mary Dupre is kneeling in the center.

jugs, chamber pots, milkpans, porringers, cider or ale mugs, and possibly flowerpots. It appears that Hazeltine continued with these very traditional forms until his death in 1880.

Oral tradition placed Hazeltine's shop on the edge of a field just north of Hopkinton Road, and a small earth mound appeared to be the shop location. Mary Dupre and I began digging there in the spring of 1982, and our work lasted through the summer of 1984 (fig. 5.31). We discovered extensive "waster" dumps on both sides of Hopkinton Road, and then we located the foundations of two kiln houses that had waster sherds, kiln furniture, and bricks densely packed across the floor of each structure. The remains of each kiln house consisted of dry-laid stone foundation walls piled two or three stones high, and a substantial stone kiln base was located in the center of each building. We completely exposed an oval kiln base inside what we called "kiln house I," and then we partially exposed a square or rectangular kiln base in "kiln house II" (figs. 5.32 and 5.33). Just to the south we tried to locate Hazeltine's pot shop, and we actually did find a line of foundation stones and a thick lens of gray clay that may have been its floor.

At the site we found large quantities of kiln furniture, including hundreds of setting bars, stilts, wedges, and stackers, many of which had been broken during use as they separated vessels inside the kilns during firing (fig. 5.34). While we didn't find any pottery-making tools, we recovered great quantities of wasters, kiln furniture, and products (fig. 5.35), and these provided much useful information about some of the technological problems dealt with in this family-based industry (see Starbuck and Dupre 1985a, 1985b, Dupre 1985b).

It appears that there was constant demand for all of these pottery vessel forms throughout the ninteenth century, but whitewares eventually would replace them. By the late 1800s, redware potters like Joseph Hazeltine were producing flowerpots and little else.

FIG. 5.32. Foundations of kiln house I (foreground) and kiln house II (rear) at the Joseph Hazeltine pottery site.

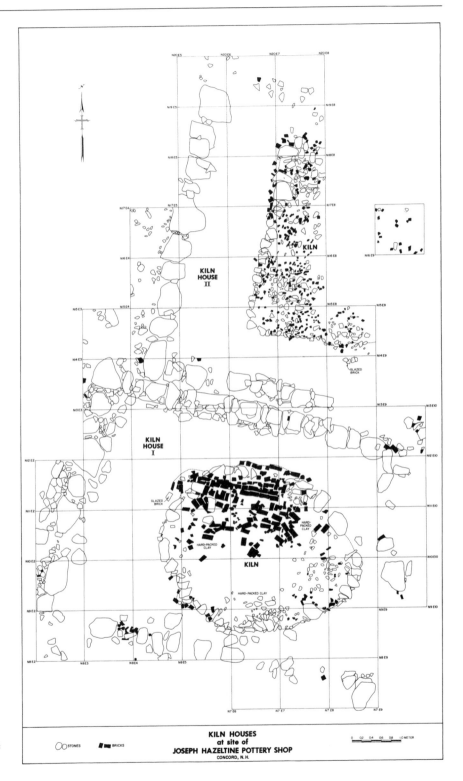

FIG. 5.33. Foundations of the two kiln houses excavated at the Joseph Hazeltine pottery site.

FIG. 5.34. Examples of wedges (top) and stilts (bottom) found in the dumps at the Joseph Hazeltine pottery site. The wedges were used to level vessels inside the kiln, and the stilts were used as supports for individual vessels.

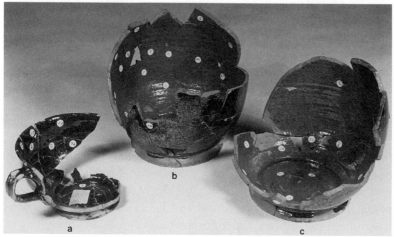

FIG. 5.35. Redware vessels excavated at the Joseph Hazeltine pottery site: *a* is a porringer, and *b* and *c* are jugs. The porringer has a dark-brown glaze on both the exterior and interior, while the jugs have a black glaze on the exterior and a clear glaze on the interior.

A Spectacular Governor's Plantation

"Wentworth House," located within what is now Wentworth State Forest near Wolfeboro, was possibly the largest, grandest mansion in New Hampshire on the eve of the American Revolution. It was the centerpiece of an estate of more than three thousand acres, referred to even today as "America's Oldest Summer Place." While the term "plantation" is usually associated with the American South, New Hampshire's last colonial governor, John Wentworth, created a summer home in Wolfeboro between 1768 and 1775 that was on the scale of a plantation and was intended to open up the interior of New Hampshire to settlement. Because he was a loyalist he was forced to flee to Canada before his grand estate could be completed.

Wentworth House, its outbuildings, and the surrounding estate were confiscated by patriots, and Governor Wentworth was never allowed to return. The great mansion burned down in 1820 while in other hands, and the property was then abandoned. The 1820 owner had made the mistake of throwing some dry shingles into a fireplace in the chilly morning, and they were sucked up the chimney and scattered across the roof. The mansion, the centerpiece of a governor's vision for the future of New Hampshire, was gone within a few hours.

Forest grew over the ruins, and little happened there for the next 165 years, except for the small excavation by the Civilian Conservation Corps in 1934 and 1935, when they placed a drain in the bottom of the mansion cellar hole. It was Dennis Chesley, then at the University of New Hampshire, who told me about this wonderful rural estate in 1981. My first visit to the site was in 1982, and after a picnic on the grass with a friend, I was treated to a case of poison ivy so virulent that I had to go to the emergency room of Concord Hospital. I was but the first of many to have this experience as our research proceeded!

After a lengthy planning period, Mary Dupre, Gary Hume, and I finally commenced the excavation of the Governor Wentworth Plantation in 1985, and our dig lasted through 1988. Our four seasons of excavation were sponsored by Plymouth State College, the New Hampshire Division of Parks and Recreation, and the Division of Historical Resources. Because the property had been abandoned so early and had been left undisturbed, it was an ideal site for an archeological investigation. Governor Wentworth had been the last of New Hampshire's provincial governors, a man of vision, a builder of roads through the frontier, and he had even provided the land for Dartmouth College. This was truly an inspirational site for us to explore.

What is left of the Wentworth plantation is now dominated by the great cellar hole of Wentworth House, which is 100 feet long by 40 feet wide; it also had a kitchen wing added on to its western end (figs. 5.36 and 5.37). But the plantation layout also features the remains of several other structures. We located two sets of stables and coach houses (one connected to a large barn); a dairy and a well; a large cistern; and a small outbuilding located just southwest of the mansion. We excavated at all of these sites, but we know from historical sources that other structures once existed here, including smokehouses, an ash house, a carpenter's shop, a blacksmith's shop, a sawmill, a caretaker's house, and houses for other workers. Clearly much more archeology could be done here in the future.

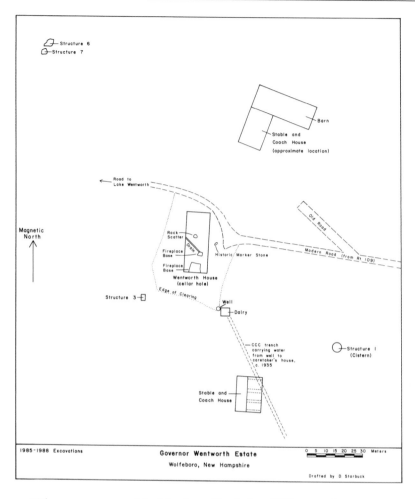

Structure 6
Structure 7

Barn

Stable and
Coach House
(approximate location)

Road to
Lake Wentworth

Old Road

Magnetic
North

Rock
Scatter

Fireplace
Base

Drain

Historic Marker Stone

Modern Road (from Rt 109)

Fireplace
Base

Wentworth House
(cellar hole)

Edge of Clearing

Structure 3

Well

Dairy

CCC trench
carrying water
from well to
caretaker's house,
c. 1935

Structure 1
(Cistern)

Stable and
Coach House

1985-1988 Excavations

Governor Wentworth Estate
Wolfeboro, New Hampshire

0 5 10 15 20 25 30 Meters

Drafted by D Starbuck

FIG. 5.36. Modern site map of the Wentworth plantation in Wolfeboro.

When we excavated inside the cellar hole of Wentworth House, we discovered that the burning of the mansion left a broad exposure of burnt boards, floor joists, and sheets of crumbling plaster throughout the cellar (fig. 5.38). The bases of two large fireplaces were surrounded by sherds of creamware, pearlware, and redware, together with hinges, coins, knives, a pewter spoon, and thousands of nails and fragments of window glass. The house may never have been truly "finished" before the governor left, but artifacts and architectural remains are everywhere even today. It is rather sad that this lightly visited site is now more of a "parking spot" for local couples than it is the former summer home of a revered governor. On the very first day of our dig in 1985, as we began clearing brush from the cellar hole, we discovered a "sexual aid" that had no doubt been thrown from a parked car. After that, we always referred to it as "artifact #1 from Wentworth House," but propriety dictates that I not mention what it was!

From an archeological standpoint, however, it is the outbuildings—not the mansion—that are the richest part of Wentworth plantation. The feature that we called "structure 1" is a deep, circular foundation, probably a cistern, that is about 250 feet east-southeast of Wentworth House (figs. 5.39 and 5.40). At the very bottom, about six feet down, we discovered a mass of mid- to late-eighteenth-century dishes and wine bottles (figs. 5.41, 5.42, and 5.43). Pieces were missing from some of the

FIG. 5.37. A plan view of the great cellar hole of Wentworth House, showing the pits excavated between 1985 and 1988.

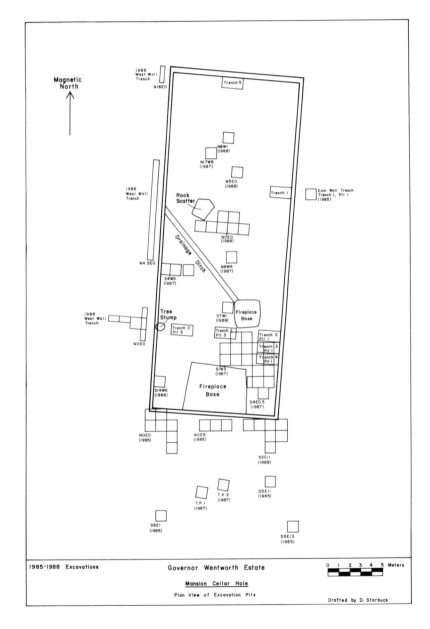

FIG. 5.38. The excavation inside the cellar hole of Wentworth House.

FIG. 5.39. The surface of structure 1 (a cistern) at the Wentworth plantation.

FIG. 5.40. Structure 1 at the completion of excavation in 1986. Dishes and bottles rested on the clay at the bottom of the cistern.

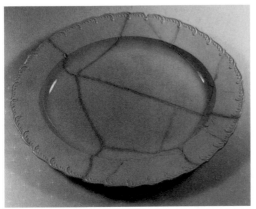

FIG. 5.41. A creamware platter, with feather-edged rim, discovered inside structure 1 at the Wentworth plantation (oval, 15 by 18½ inches).

FIG. 5.42. A creamware sauce boat, with feather-edged rim and a rope-twist handle, from structure 1 at the Wentworth plantation (7 inches long, 5¾ inches high).

vessels, but many of the plates were stacked on top of one another, giving the impression that they had been carefully placed there all at once. I fondly remember how tourists would visit as I was uncovering the rich cache of ceramics in the bottom of the cistern, and I would sometimes shift my position just enough to "give them a peek"—just to whet their curiosity! Archeologists are most definitely aware of the excitement that can be generated by the act of discovery, and some of us love "performing" when we have an attentive audience that enjoys watching us troweling and brushing!

Of the vessels we discovered inside the cistern, creamware was the most common type of ware, with white salt-glazed stoneware a close second. The assemblage dated no later than the 1770s, so this cache of unusually fine pieces definitely dated to the period of the governor's residency. We can only wonder whether he missed his fine dishes once he had departed for Canada! Soon after we glued these dishes back together, we mounted an exhibit of some of the best of them that went on display in the New Hampshire State House in Concord. New Hampshire's then-governor, John Sununu, was present for the opening, and it was a thrill to watch a second governor—after an interval of about 211 years—actually handle the same dishes!

The other outbuilding that was unusually rich had a six-foot-deep cellar hole (fig. 5.44). It was located 77 feet southwest of the mansion, and we identified it as "structure 3." It was richly filled with an assortment of ceramics and glass that dated principally to the time of the plantation's abandonment in 1820. This assemblage included many plates, cups and bowls of English creamware and pearlware (figs. 5.45

FIG. 5.43. A white salt-glazed stoneware chamber pot found inside structure 1 at the Wentworth plantation (7½ inches in diameter, 5¼ inches high).

FIG. 5.44. The cellar hole of structure 3 at the Wentworth plantation after excavation was completed in 1987.

FIG. 5.45. A creamware fruit bowl found inside structure 3 (oval, 11½ inches by 13½ inches).

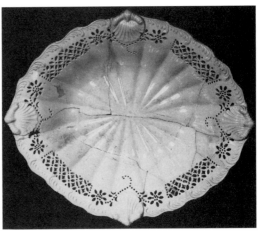

FIG. 5.46. A second creamware fruit bowl found inside structure 3 (oval, 11½ inches by 13½ inches).

FIG. 5.47. A reconstructed redware milkpan discovered inside structure 3 (17½ inches in diameter, 3¼ inches high). It has a clear lead glaze and five small "mend holes."

FIG. 5.48. A spectacular case bottle found in structure 3 (19 inches tall). The reconstruction, by SCRAP volunteers, was little short of miraculous!

and 5.46), sherds from 16 redware milkpans (fig. 5.47; consequently we know that much of this was a dairying assemblage), several redware jars and chamber pots, and two 19-inch-high glass case bottles (fig. 5.48). Most of the creamware was fairly late in date and, combined with a preponderance of pearlware, was indicative of a reasonably well-to-do family of the early nineteenth century. But these ceramics did not approach the quality of the governor's own pottery, and most of them fall too late in time to have been his.

No other excavation in northern New England has ever produced as much evidence for the layout of an eighteenth-century formal estate, and the Governor Wentworth plantation stands out as one of the most exciting and significant survivals from the colonial period in New Hampshire (see Starbuck 1988b, 1988c, 1989). There are no current plans for additional excavations there, but locating the missing workers' houses and craft shops would help bring completion to this very interesting project.

Other Categories of Historical Sites

Many other categories of historical sites have been excavated in New Hampshire. For example, Faith Harrington initiated work in 1986 on the Isles of Shoals, nine rocky islands that are located about eight miles off the coast of New Hampshire and Maine. In the seventeenth and eighteenth centuries there was a very active cod fishery there, and Harrington used archeology to show "its transition from a temporary fishing station to a permanent settlement" (Harrington 1992: 249). This project—sponsored by EARTHWATCH, the University of Southern Maine, and others—was quite unique in that no other archeologist had systematically focused on understanding early fishing sites and changing settlement practices just off the coast of New Hampshire (also see Harrington 1985b, 1985c; Harrington and Kenyon 1987; Kelso and Harrington 1989). Her work on Appledore Island, in particular, focused on discovering sheds and living quarters from the early fishery.

Other examples of historical digs in New Hampshire have included the excavation by Alaric Faulkner (now at the University of Maine in Orono) of the site of a brickyard in Nelson; Billee Hoornbeek (formerly of the University of New Hampshire) digging the Old Parsonage site in Newington in 1978 and 1979; and Robert Ewing (formerly of the University of New Hampshire) leading efforts in 1979 and 1980 to discover the sites of mid- or late-eighteenth-century pesthouses in Exeter and at the Newmarket Town Farm (Anonymous 1980–1981). Pesthouses were used to isolate the victims of contagious diseases, especially smallpox, from the general population. Neither site revealed any evidence for medicines that might have been used for treating disease, so the field work at these two sites ultimately proved to be inconclusive—we do not know whether either site was actually a pesthouse.

Perhaps the most frequent category of historic site excavation in New Hampshire could be called "backyard archeology." Backyards tend to be far richer than front yards, chiefly because many more activities occurred in the rear of the main house. It was invariably in the rear, out of sight, where privies, trash pits, outbuildings, and most open-air activities were located. The backyards of many of New Hampshire's older houses have been dug by archeologists, and this is an excellent way to study the changing fortunes and consumption practices of the residents of any historic house.

FIG. 5.49. The southwest corner of the Abiathar Britton house in Orford. The 1998–1999 excavation is in the foreground, next to what had formerly been a brick carriage house (now replaced with a modern addition).

FIG. 5.50. A stone foundation discovered behind the Abiathar Britton house.

One of the more recent of these "backyard excavations" in New Hampshire was one I conducted in 1998 and 1999 at the site of a trash-filled cistern or cellar at the Squire Abiathar Britton house in Orford. The Abiathar Britton house is on Main Street in this small, prosperous town on the upper Connecticut River. As is so often the case, this feature was discovered purely by chance by the current property owners, Karl and Rika Schmidt, as they tried to prevent water seepage into their cellar. They found an extensive trash deposit just beyond the rear wall of what had originally been an attached brick carriage house (fig. 5.49).

After I was contacted by the Schmidts, I led a team from Plymouth State that exposed a four-by-five-foot stone-lined feature that was nearly six feet deep just behind the rear door of the house (fig. 5.50). Inside this foundation we discovered an incredible collection of ceramics and glass that had most likely been inside the Britton house at the time of an 1830 fire. The burned late-eighteenth- and early-nineteenth-century contents of the house that were thrown into the backyard cistern provide evidence for a wealthy family that owned many heirlooms, including a soapstone inkwell, two English-made chamber pots with "Scratch Blue" decoration (figs. 5.51 and 5.52), storage pots, serving bowls, tablewares, and two Portobello pitchers (manufactured in Scotland), decorated with Greek chariots and horses (fig. 5.53).

Our finds also included several pearlware teacups that were decorated with pineapples, long a symbol of hospitality (fig. 5.54). We even found glass tumblers and a snuff bottle (fig. 5.55) buried inside this small foundation. If the wealth of a prominent New Hampshire family is best shown through its trash, then the Abiathar Britton family was very well off indeed. The "heirloom" pieces that had once been inside their house suggest that they were really collectors of ceramics and not simply consumers. It also appears that they engaged in a great deal of entertaining, given the large number of serving vessels found inside the cistern. After the 1830 fire, the Britton family built the house that currently stands on the property and, no doubt, furnished it with all-new dishes (Starbuck 2000).

FIG. 5.51. Two white salt-glazed stoneware chamber pots discovered inside the trash-filled foundation behind the Abiathar Britton house. The one on the left has a molded King George III medallion (8 inches in diameter, 5¼ inches high), and the one on the right has a molded American Eagle medallion, a simplified form of the Great Seal of the United States (7⅞ inches in diameter, 5½ inches high).

FIG. 5.52. The two stoneware chamber pots, showing the King George medallion (left) and the American Eagle medallion. Drawing by Ellen Pawelczak.

FIG. 5.53. A Portobello pitcher found in the foundation behind the Abiathar Britton house (5 inches in diameter, 5½ inches high). This beautiful transfer-printed pitcher, manufactured in the late 1700s, would have been made by the Scott brothers near Edinburgh, Scotland.

FIG. 5.54. Pearlware teacups handpainted with pineapple decoration and tea leaves, discovered in the foundation behind the Abiathar Britton house (each 3¼ inches in diameter, 1⅞ inches high).

FIG. 5.55. Glass snuff bottle found at the very bottom of the foundation behind the Abiathar Britton house (1¼ inches in diameter, 3½ inches high). It reads: "BY THE KING'S PATENT[:] TRUE CEPHALICK SNUFF."

What Does the Future Hold?

New Hampshire historical archeology has made tremendous advances over the past 40 years, and at least half of all digs in the state are now conducted at the historic period sites of European Americans. Research will hopefully continue into sites and topics that will further reveal the essential character of historical New Hampshire, its customs and its values. Much remains to be done, and there have regrettably been very few archeological studies of minority groups in the state, perhaps reflecting New Hampshire's relatively homogeneous population. Also, archeology has rarely been used to say much about women's activities, and there have been few efforts to locate the remains of early posthole houses and other forms of poorly known architecture that were holdovers from English medieval styles. Also, relatively little has been published about farm sites and not enough about early industries.

Nevertheless, New Hampshire historical archeology is increasingly well known beyond the borders of the state, and publications about the archeology at Strawbery Banke and Canterbury Shaker Village have often reached a national audience. While New Hampshire's historic archeological sites may not be quite as well known as those in Massachusetts or Virginia, many sites have been little disturbed by modern development. Thus it is the integrity and excellent preservation of New Hampshire's archeological sites that are our greatest assets.

Chapter 6

Industrial Archeology

N EW HAMPSHIRE has an exceptional number of early industrial sites that have maintained their integrity down to the present day. Many of these sites are truly one of a kind and thus uniquely able to provide insights into how industry has operated and grown within this state. Few states have had as rich an industrial heritage as New Hampshire, but it is no accident that there have been so many industrial survivals. After all, New Hampshire has always had a diversified industrial base, and modern growth—and concurrent site destruction—has been largely restricted to the southern and southeastern portions of the state. (See the 1994 issue of *IA: The Journal of the Society for Industrial Archeology* for an excellent collection of articles documenting some of the best industrial survivals in the state of New Hampshire.)

Several once-important industries and workplaces have declined and moved elsewhere over the past century—notably textile manufacturing, shoemaking, and shipbuilding—and many of New Hampshire's older industries have vanished completely. However, many other industries remain viable today, including granite quarrying, lumbering, printing, and papermaking. There also are a host of "high tech" industries that have moved into the southern part of New Hampshire, although most of these are relatively ephemeral (from an archeological perspective) and do not leave many physical remains behind. Still, there is no question that companies in the electronics, computer, and software fields have totally revolutionized the ways in which New Hampshirites view themselves and their workplaces.

New Hampshire's Industrial History

The English settlers who colonized New Hampshire in the early 1600s combined fishing along the seacoast with lumbering in the interior of the state, and one of the most significant early industries involved the production of masts and spars for ships in the British Royal Navy

FIG. 6.1. Excavating the remains of the Second Family Blacksmith Shop at Canterbury Shaker Village, 1996.

(Malone 1964). But many maritime industries declined as the population moved inland, and farmers and craftsmen increasingly dominated the work force of New Hampshire. In the seventeenth and eighteenth centuries, every town had grist mills, sawmills, and wood-turning mills for processing forest products (Candee 1970); nearly every town had blacksmith shops (fig. 6.1; Starbuck 2000a) and redware potters (Dupre 1985b, Pendery 1985, Starbuck and Dupre 1985a, 1985b); ropewalks produced cordage in the Portsmouth area; the stonecutting of granite became important in Concord and Milford; and extensive brickyards developed on the interior reaches of Great Bay and along the Merrimack and Connecticut rivers. It is difficult to view most of these industries in New Hampshire today, but a fortunate exception is the Taylor Up-and-Down Sawmill in Derry, a forestry education and demonstration center that is administered by the New Hampshire Division of Forests and Lands. The Taylor Sawmill is operated for the public several times each summer, demonstrating how an early, water-powered sawmill would have functioned. (The current Taylor Sawmill can handle logs up to 10 feet in length and 28 inches in diameter.)

While adjacent states had significant quantities of iron ore and established many ironworks, New Hampshire had relatively few veins of iron, and there have been only three ironworks in the history of the entire state. The most successful of these was the New Hampshire Iron Foundry, which commenced operation in the northern town of Franconia in 1805 next to a good source of ore (Rolando 1993, Webber 1973, Welch 1972). Several small recording projects have been held at the furnace in Franconia (fig. 6.2), including work in 1991 that was directed by Duncan Wilkie using students from Plymouth State University. More recently (from 1994 to 1996), a larger project was conducted there by Victor Rolando using members of the Northern New England Chapter of the Society for Industrial Archeology (Rolando 1996; see the box "The Northern New England Chapter, Society for Industrial Archeology"). The still-standing blast furnace in Franconia was built on the east side of the Gale River in 1859, but all of the associated ironworks buildings were destroyed by fire in 1884. The blast furnace is in amazingly good condition, although there has been some deterioration of the firebrick lining and crucible, and some stones have fallen from the furnace top and from the outer walls. This is a practically unique industrial survival, because the Franconia furnace stack is eight-sided, unlike any other in New England.

Large-scale industrialization did not reach New Hampshire until early in the ninteenth century, when the ready availability of water

FIG. 6.2. The octagonal blast furnace of the Franconia Ironworks, 1994.

★ The Northern New England Chapter, Society for Industrial Archeology

The field of industrial archeology first developed in Great Britain, birthplace of the Industrial Revolution and home to an amazing variety of still-standing industrial ruins. It was in the late 1960s that this fascination with bridges, railroads, canals, textile mills, and innumerable other industries finally arrived in America, championed most notably by Robert Vogel, a curator in the Smithsonian's National Museum of American History. The American-based Society for Industrial Archeology (SIA) was born in 1971, after which it was headquartered for many years out of Vogel's office at the Smithsonian. More recently, its home has been the Department of Social Sciences at Michigan Technological University. The SIA publishes a journal (*IA*) as well as a newsletter (*SIA Newsletter*), and it hosts an annual conference in May or June as well as an annual fall tour. The emphasis has traditionally been upon documenting the architectural remains of early industries, but there is also a strong emphasis upon studying industrial processes. As such, process tours—which visit industries that are still capable of revealing how manufacturing was (and is) done—are a very popular part of this field.

The SIA has several local chapters, each of which hosts its own meetings and tours. New Hampshire is home to the Northern New England Chapter (NNEC), first organized

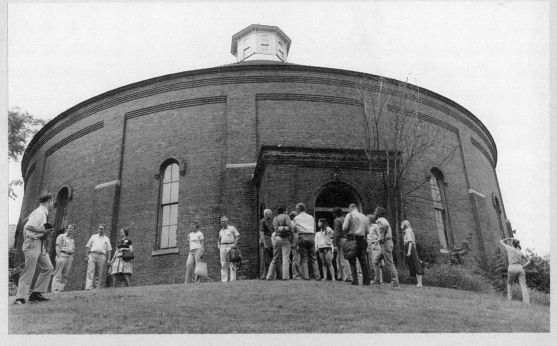

The very first industrial tour organized by the Northern New England Chapter—the Concord Gasholder House on July 26, 1980. James Garvin, New Hampshire State Architectural Historian, is standing at the far left (with camera in hand), and Christine Fonda (Rankie), New Hampshire National Register, Preservation Tax Incentives and Covenants Coordinator, stands second from the far right.

power—especially on the Merrimack, Lamprey, Piscataqua, Cocheco, and Androscoggin rivers—began to attract entrepreneurs in the textile industry and investors from the Boston area. The manufacture of textiles subsequently dominated much of New Hampshire throughout the nineteenth and early twentieth centuries, and major textile manufacturing centers developed at nearly every falls on every river in the southern part of the state. At their peak, perhaps 40 percent of the workers were French Canadian, but they were accompanied by large numbers of Greeks, Finns, and others. "Mill girls" were a significant part of this work force (Dublin 1979), but ten times as many young women were engaged in home industries as in factory production. Manchester, Nashua, Dover, Somersworth, and a host of other communities located on good transportation arteries established factories for the processing of cotton, while a great many smaller towns developed mills that relied upon wool. The first textile mill was located in the small town of New Ipswich, just north of the border with Massachusetts. Another small mill town, Harrisville, in the southwestern part of the state, has spun woolen yarn since 1794, and it has been much more successful than other communities in maintaining the integrity of its workers' housing, canals, ponds, and mill buildings (fig. 6.3; Armstrong 1970, Wingerson 1994). As a result, Harrisville was designated a national historic landmark in 1977, and it has frequently been described as the most intact early-nineteenth-century mill village in the United States. The last surviving woolen mill in

in 1980 by Christine Fonda (Rankie) and myself. The NNEC represents New Hampshire, Maine, Vermont, and eastern New York State and sponsors spring and fall meetings, occasional hands-on recording projects, and an annual conference held on a weekend in February (and alternating between Plymouth State University one year and a southern New England site the next). Over the past 26 years, at least half of the NNEC's meetings have been held in New Hampshire (reflecting New Hampshire's central location within the region being served), and this has opened up an impressive number of early (and modern) industrial sites to visitation by scholars, buffs, and those who simply can't resist old industry!

Many of the industrial sites described in this chapter have been visited by the NNEC, including the Concord Gasholder House, Frye's Measure Mill, Harrisville, the Page Belting Company, the Sewall's Falls Dam, the Franconia Ironworks, Canterbury Shaker Village, the Belknap Mill, the Simplex Wire and Cable Company, and others. As with the national membership of the SIA, most Northern New England Chapter members are not archeologists who dig. Rather, they come from every possible profession, including historic preservation, architectural history, engineering, various positions in the museum field, and even modern manufacturing. But what they all share is a love for old industry and especially for the buildings and machinery that have survived down to the present day. New Hampshire has so many great industrial remains that this state is truly an industrial archeologist's paradise!

FIG. 6.3. The Granite or Lower Mill (right) in Harrisville, 1984.

Harrisville closed in 1970, but in 1971 Harrisville Designs was created, and this small, family-owned business has continued to produce high quality woolen yarns down to the present day.

The larger textile complexes ultimately failed, however, including such giants as the Amoskeag Mills in Manchester and the Sawyer Woolen Mills in Dover, as production moved to the South in the 1920s and 1930s. The failure of the New England textile industry to use profits to modernize production has been well documented elsewhere (Gross 1988), and too many New Hampshire textile manufacturers were no different from their short-sighted competitors in southern New England. After textile production moved out, there was an intermediate stage when some of the brick structures were used for shoe manufacturing, but even most of the shoe companies, such as Nike, have moved their production out of New Hampshire.

Memories of the textile industry persist, of course, because the Amoskeag Manufacturing Company had become the largest textile company in the world prior to its closure in 1935 (Hareven and Langenback 1978, Mayer 1994). Southern New Hampshire still has abundant brick textile mill buildings that have been converted into housing, small businesses, restaurants, and shops. Many of the buildings of the Great Falls Manufacturing Company, the Salmon Falls Manufacturing Company, the Cocheco Manufacturing Company, the Nashua Manufacturing Company, the Sawyer Mills, and the Newmarket Mills continue to be viable workplaces or residences within their present-day communities, but their uses are no longer industrial. The Belknap Mill survives as an excellent small museum (Boswell 1994); part of the Amoskeag Mills is now occupied by the University of New Hampshire at Manchester; the Manchester Historic Association's Millyard Museum houses artifacts that came from the same buildings and era as the museum; and many other textile mills have been converted into housing.

Still, one of the hallmarks of New Hampshire industry has been diversity, and the lumber, pulp, and paper industries have consistently been strong, as exemplified by the Monadnock Paper Mills on the Contoocook River in Bennington, where high-performance graphic arts papers are made. Maritime industries continue to be successful in the Portsmouth area, including the Simplex Wire and Cable Company in Newington (taken over by Tyco in 1974), which is the sole manufacturer of transoceanic undersea cables in the western hemisphere. Elsewhere in the state, the largest bookmatch company in the world, D. D. Bean and Sons Co., is based in Jaffrey. Small woodworking mills still make specialty products to order, and preeminent among these is the water-powered Frye's Measure Mill in Wilton (figs. 6.4 and 6.5). Frye's Mill has made woodenware since 1858, including colonial and Shaker-style boxes, measures, and piggins. Many of the older industries have *not* survived, however, including glassmaking, mining, some types of quarrying (such as soapstone), coachmaking, blacksmithing, ironmaking, tanneries, and baseball production. The old industries have given way to skiing, tourism, arts and crafts, and the manufacture of Annalee Dolls (in the town of Meredith). For every "old" industry that has vanished, a new one has arisen to take its place.

Several early efforts at glassmaking occurred in New Hampshire, with the first occurring in the town of Temple (1780–1782), which was followed by two glass factories in Keene (1814–1855 and 1815–1841) and glassworks in Suncook (1839–1850), Stoddard (1842–1873), and South Lyndeborough (1866–1886). None of these efforts proved particularly successful, but products have survived from each, and some of the factories have left substantial ruins behind.

Originally about four hundred covered bridges stood in New Hampshire, and 54 of these have survived and are administered by the New Hampshire Department of Transportation. The most impressive of all is certainly the 1866 Cornish-Windsor covered bridge across the Connecticut River between Cornish, New Hampshire, and Windsor, Vermont (figs. 6.6 and 6.7; Jackson 1988: 82–83, 98, Keyworth 1973, Lewandoski 1990). The Cornish-Windsor bridge features a Town lattice truss, has two spans, and is 449 feet 5 inches long and exactly 24 feet wide. This is unquestionably one of the most remarkable still-in-use bridges in the world, and it is the longest wooden covered bridge in the United States. But New Hampshire also contains many other historic bridge types,

FIG. 6.4. Exterior of Frye's
Mill in Wilton, 1984.

FIG. 6.5. Interior of Frye's
Mill, 1982. Various products
are made to order at different
stations across the shop floor.
Shaker boxes are stacked at
the right.

FIG. 6.6. The Cornish-Windsor Covered Bridge across the Connecticut River, 1980.

FIG. 6.7. The Cornish-Windsor Covered Bridge as it underwent restoration in 1988.

some of which were inventoried by the Historic American Engineering Record during their New England inventory in 1972 and 1973 (*New England: An Inventory of Historic Engineering and Industrial Sites*, 1974). Other bridges have been included by Donald Jackson in his National Trust guide to bridges and dams (1988), and just one, the Monadnock Mills Bridge (1870) in Claremont, was included by Eric DeLony in his survey (1993) of cast- and wrought-iron bridges in America.

New Hampshire's Industrial Archeology

There are many outstanding examples of New Hampshire's well-preserved industrial sites that have been documented by industrial archeologists. These include an 1888 gasholder house in Concord, which is the most intact gasholder in the United States (figs. 6.8 and 6.9)—here gas manufactured from coal was stored before being distributed to the city (Taylor 1984); the site of the Nichols-Colby up-and-down sawmill in Bow, which was built in the early nineteenth century and destroyed in the hurricane of 1938—a replica of this was later constructed at Old Sturbridge Village (Penn and Parks 1975); the Page Belting Company of Concord, begun in 1868, which was one of the last manufacturers of flat power transmission leather belting in the United States and a world leader in many types of leather products (fig. 6.10; Howe 1993); the longest timber crib dam in the eastern United States (Starbuck 1990c);

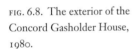

FIG. 6.8. The exterior of the Concord Gasholder House, 1980.

FIG. 6.9. The interior of the Concord Gasholder House, 1980, with visitors standing on the actual gasholder tank.

low-head mill sites where wood, grain, and cloth were once processed at Canterbury Shaker Village (Starbuck 1986b, 2004); and the Brown Company (now the James River Corporation), which was the center of a huge forest products industry in the northern industrial city of Berlin (Taylor 1993).

Syntheses of New Hampshire's industrial archeology have appeared in the journals *Historical New Hampshire* (Starbuck 1985b), *The New Hampshire Archeologist* (Howe 1994a), and *IA: The Journal of the Society for Industrial Archeology* (Starbuck 1994a). Also, extremely useful books have been published on New Hampshire's textile companies and communities (Hareven and Langenback 1978, Armstrong 1970), its maritime industries along the seacoast (Candee 1985), its hand-tool makers (Garvin and Garvin 1985), its glassmaking (Starbuck 1986a), its logging railroads (Belcher 1980), and its bridges (Keyworth 1973, Openo 1988). Many of New Hampshire's industrial sites are thus well known and accessible through a host of publications.

However, in addition to the more famous sites like Harrisville and the Amoskeag Mills, there are a great many other industrial sites in New

FIG. 6.10. Interior of the Page Belting Company in its original location in Concord, 1983. In this demonstration, strips of leather are being lapped and glued together to form power belting.

Hampshire that have rarely received the national recognition that they deserve. These include the Abbot-Downing Company, makers of the horse-drawn Concord Coach, which peaked about 1890 and went out of business in 1915 (Wilding-White 1994); the Mount Washington Cog Railway (Starbuck 1994b, 2003–2004); the Draper-Maynard Sporting Goods Company of Plymouth and Ashland, manufacturers of baseballs, gloves, and other sporting goods between 1840 and 1937 (Freeman and Donahue 1994); the Sewall's Falls hydroelectric facility, which went into operation in 1894 and was one of the first commercial polyphase hydro-electric stations in the United States (Starbuck 1990c; Howe 1994b); the Swenson Granite Works in Concord, founded by John Swenson in 1883, which, with its affiliates, is now the largest quarrier of granite in North America (D. Garvin 1994); and a number of brickmaking locations, the last of which closed in 1994 (the Kane-Gonic Brick Corporation in Rochester; J. Garvin 1994).

The Glasshouse at the New England Glassworks

While most industrial archeology does not involve digging, Boston University's excavations at the New England Glassworks in Temple easily comprised the largest industrial dig ever conducted in the state of

New Hampshire. This glass factory was the first to be built north of Boston, and the entrepreneur who built the factory, Robert Hewes, once wrote that he had glassmakers in his employ even before he went to Temple. This remote location was perhaps chosen because of the easy availability of fuel, sand, and potash, but markets for the finished glass were well to the south and along poor roads. The factory was in production for only two years (1780–1782), and afterward both the glasshouse and the workers' village (see the previous chapter) were permanently abandoned. The forest grew over the site of the old glassworks, and this was followed by two centuries of bottle collectors, loggers, and hunters wandering through the ruins.

In 1974, James Wiseman of Boston University (BU) selected the glassworks as a long-term base for college field schools, and Frederick Gorman (also of BU) and I directed digs there between 1975 and 1978. In the four years of digging, we excavated and mapped the ruins of the glasshouse stone by stone (fig. 6.11), and in the process we tried to determine what types of glass were being made there, how standardized the products were, and how the glasshouse was actually constructed. Did the glasshouse and its furnace have a relatively new design of Hewes's creation, or had he adopted one of the standard European designs of the period? Either way, history tells us that the factory produced crown window glass—the first made in America—as well as "junk" bottles (fig. 6.12) and "vessels suitable for use in chemistry" (Hewes 1781). Each crown of window glass would have begun as a heavy gather of glass on the end of a blowpipe, and this was spun in the air until centrifugal force modified it into a large flat disk, ready to be cut into panes.

As our work progressed, we discovered that the ruins of the glasshouse measured 69.5 by 67.6 feet (the building was probably 65 by 65 feet when newly built); it contained a massive central furnace of German design (fig. 6.13); there were several annealing and calcining ovens in the northern and eastern walls of the factory (fig. 6.14); and everywhere there were thousands of fragments of unfinished glass, cullet, and crucibles. The crucibles had held the molten glass inside the furnace, and the intense heat had split the crucibles apart. We even found a pewter "USA" button (see fig. 5.13) and a single pair of gold-leaf sleeve links, both of which were lying in the furnace area. Altogether we found some 202,726 fragments of glass (fig. 6.15), and over 90 percent of the excavated glass was located on the working floor of the glasshouse. After many, many years of collectors combing through the ruins looking for glass samples, it is amazing that we found as much as we did.

If the number of bubbles and inclusions in the glass are any indication, then product quality may have been low, but no doubt the best pieces were sold and are no longer at the site. The huge quantities of

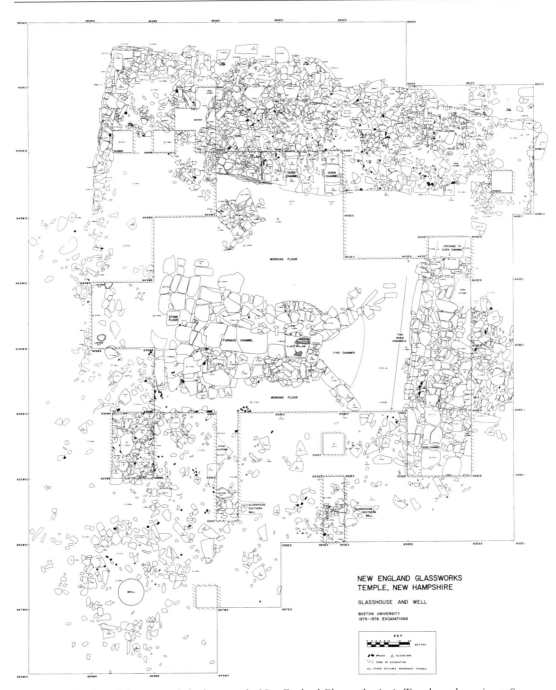

FIG. 6.11. A plan view of the excavated glasshouse at the New England Glassworks site in Temple, as drawn in 1978.

FIG. 6.12. A light-green bottle in the collection of the Temple Historical Society, reported to have been made at the New England Glassworks (height: 9½ inches).

FIG. 6.13. The excavation of the furnace foundation inside the glasshouse of the New England Glassworks, 1976.

FIG. 6.14. Two annealing ovens excavated inside the north wall of the glasshouse at the New England Glassworks, 1978. The scale boards are marked in 10-centimeter units.

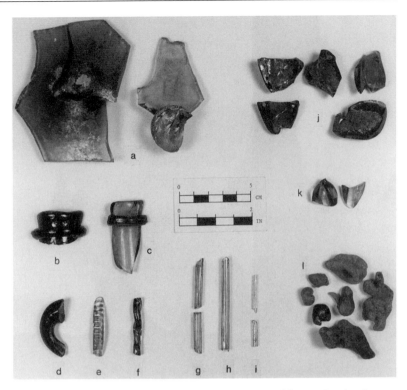

FIG. 6.15. Examples of glass products and glass waste excavated from within the glass-house at the New England Glassworks. These are: (a) bull's-eyes, the center of a sheet of crown window glass; (b) a dark-green bottle rim; (c) a light-green bottle rim; (d) a light-blue handle fragment, probably from a bowl or small pitcher; (e) a light-green, ribbed handle fragment; (f) a dark-green, twisted stem or rod; (g–i) glass tubing; (j) fragments of moil, the glass left on the end of the blowpipe or pontil when the bottle or plate was cracked off; (k) "pinched" examples of glass waste, showing the marks of the glassblow-er's pincers; and (l) examples of irregular glass spillage (found scattered everywhere).

glass waste suggest that the factory had at least some success. No doubt much experimentation was taking place, and this really was the only factory making glass in New England between 1780 and 1782. The factory in Temple was, in fact, the very first glass factory in the new United States. Temple definitely was the site of a very bold adventure in early American industry.

Most glass factories in early America were located in urban areas, guaranteeing that they would be built over and destroyed by later industries. The glass factory in Temple was in such a remote location, far from markets and skilled workers, that its failure was almost assured. Still, its isolation has helped to preserve the factory's ruins, and its very brief lifespan allows us to look at a single moment in glassmaking history.

Industry at Canterbury Shaker Village

A comprehensive archeological survey and excavations at Canterbury Shaker Village have helped to reveal how critical industry was to the Shakers and how quickly they accepted technological change in their craft manufactures and light industrial pursuits. When I began work there in 1978, I learned that a series of eight manmade millponds had been excavated and dammed along the eastern side of Shaker Village, beginning in 1800. As water flowed through the system from north to south, it powered many successive mills. The Canterbury Shakers had as many as 18 mills powered by water, and Shaker millers would release the water so as to power waterwheels and turbines before it was allowed to flow on to the next millpond and mill in the system.

It was Eldress Bertha Lindsay who first told me about the wonderful mill system that the Shakers were so very proud of, and I resolved right away to clear the brush, photograph, draw, and excavate the many mill features that had been created in the 1800s and that had gradually been abandoned throughout the 1900s. The Shakers had lacked naturally occurring bodies of water, so they created an extensive system of reservoirs, linked by ditching and supplied with water deriving from natural ponds and marshes two miles to the north. As the water slowly flowed downhill, following natural contours, it was collected in millponds and then released through raceways so as to power the mills at the south end of each pond. Earth and stone dams held water in the ponds, and stone-lined spillways (overflows) built into the dams controlled the level of water in each pond (figs. 6.16 and 6.17). Wood or metal trash racks filtered out branches and leaves that would otherwise have entered the penstocks or headraces and clogged the turbines or waterwheels (fig. 6.18).

By the time I began work in Canterbury, only one individual remained who had seen many aspects of the mill system in operation, and that was Mildred Wells. Mildred had lived with the Canterbury Shakers for practically her entire life, and because she had never formally joined the Shakers, she was free to wander at will through the mills and around the millponds. Shaker sisters typically spent very little time around the mills, because those activities were reserved for the Shaker brothers, and sisters were not allowed to venture into male areas (except to do an annual spring cleaning of the interior of the mills). Mildred liked to describe how she would walk across the muddy bottoms of the ponds on snowshoes whenever the ponds lost their water. Now that must have been an interesting sight!

FIG. 6.16. A spillway built into one of the dams on Turning Mill Pond at Canterbury Shaker Village, after the removal of beaver debris in 1980.

The Shakers' mills included sawmills (for sawing boards from logs), wood mills (for cutting up firewood), turning mills (producing brooms and chair and table legs), carding mills, grist mills, a fulling mill, a tannery, a clothiers' mill, a pump mill, and a threshing mill. We cleared the brush from atop all of these, mapped them and sometimes excavated them, and summarized the history, physical dimensions, and present condition of every last industrial feature (Starbuck 1986b, 2004). We even prepared comprehensive plans and cross-sections showing exactly how water had flowed through the Shaker mill system (fig. 6.19). All around each mill site we found abandoned equipment, machine parts, and spent fuel littering the ground. The amount of time we devoted to recording these rural industries was prodigious but absolutely necessary to show how the Shakers had modified their landscape so as to maximize power from a very low-head system.

The mill features mapped in Canterbury are repeated a dozen times over in every rural community in New Hampshire, and industrial archeologists have thus far recorded only a very few of these systems. For the Shakers, this was very simple technology, a necessary way to address

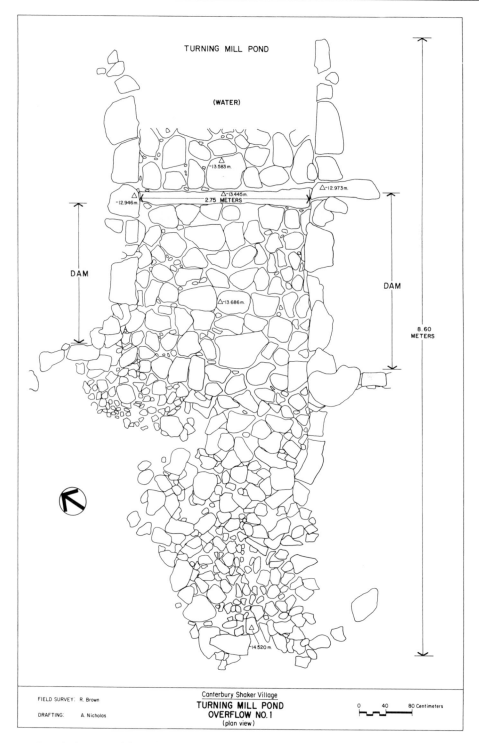

TURNING MILL POND

(WATER)

△ -13.583 m.

△ -12.946 m. △ -13.445 m. △ -12.973 m.
2.75 METERS

DAM

DAM

△ -13.686 m.

8.60
METERS

△ -14.520 m.

FIELD SURVEY: R. Brown

DRAFTING: A. Nicholas

Canterbury Shaker Village
TURNING MILL POND
OVERFLOW NO.1
(plan view)

0   40   80 Centimeters

FIG. 6.17. A plan view of the spillway depicted in figure 6.16.

FIG. 6.18. A wood trash rack and sluice gate built for use with a wood mill (in operation from 1915 to 1952) that formerly sat behind this dam on Factory Pond at Canterbury Shaker Village, 1978.

their need for power. However, the Shakers managed their power needs in a more integrated way than their neighbors, pooling resources and achieving sufficient power where formerly there had been only horse-powered mills. Unfortunately, as the Shaker brothers died out and stopped maintaining their dams and ponds on an annual basis, the dams developed leaks and the ponds lost their water. Years later, archeological mapping became necessary to document what had once been a very efficient system.

The Sewall's Falls Dam

Timber crib dams typically consist of hundreds of thousands of stones that have been hand-packed into a squared timber cribbing that is strong enough to withstand floods and ice rafting. New Hampshire does not lack for early dams on its many rivers, but one dam in particular stands out: the Sewall's Falls Dam across the Merrimack River in Concord, New Hampshire (figs. 6.20 and 6.21). Constructed between 1892 and 1894 at a cost of $125,000, this was the longest timber crib dam in the eastern United States, and it is located at the eastern end of Second Street in Concord. When built, it was 633 feet long, of which 497 feet comprised the timber-cribbed spillway and the remainder the masonry abutments that anchor the ends of the dam to the riverbanks. The Sewall's Falls Dam was constructed so as to divert part of the flow of the Merrimack

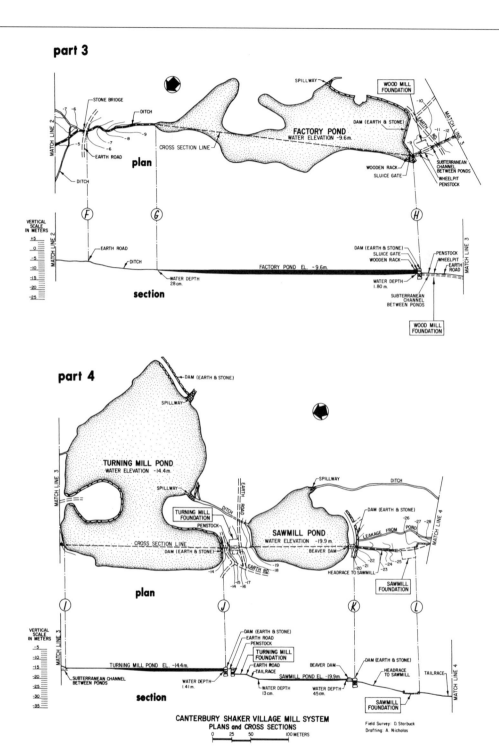

FIG. 6.19. A plan and cross-section view showing a small part of the mill system at Canterbury Shaker Village. Water slowly flowed through the system from left to right (north to south).

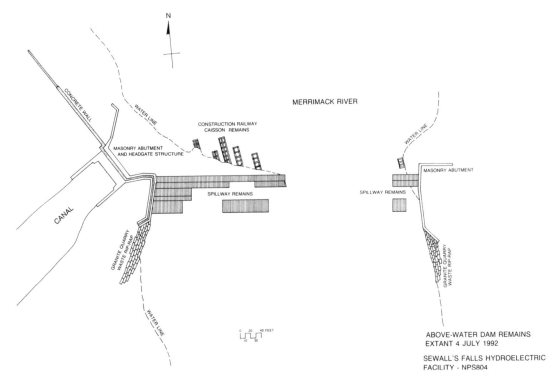

FIG. 6.20. A plan view of the now-breached Sewall's Falls Dam on the Merrimack River. North is at the top, and the river flows from north to south. *Courtesy of Dennis E. Howe.*

River into a power canal, and the remaining water flowed over the top of the cribbing (the spillway) and downstream (see Starbuck 1990c).

When construction began in late 1892 the coffer dam washed out, and the timber crib dam could not be completed until the spring of 1893. A canal and powerhouse (No. 2 Station) were then constructed and furnished with turbines, shafting, and electrical equipment. The original owner was the Concord Land and Water Power Company, which began supplying three-phase electric current over a three-phase line on September 29, 1893. This was just the second such use of a multiphase system of electric power distribution in the United States.

Floods often caused damage to the dam, and in late summer of each year it was necessary to do maintenance when the river level was at its lowest. The Sewall's Falls Dam finally ceased to be used for power generation in 1966, when its then-owner, the Concord Electric Company, decided that it was cheaper to purchase power from the Public Service Company of New Hampshire than to generate power from the generators in its power plant at Sewall's Falls (Howe 1994a, 1994b). At that point the annual maintenance on the dam ceased, and timbers began to wash out of the spillway. Finally, several days of heavy rain caused a

PLAN

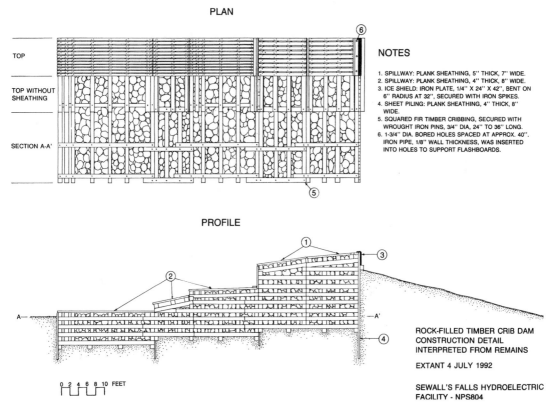

TOP

TOP WITHOUT
SHEATHING

SECTION A-A'

NOTES

1. SPILLWAY: PLANK SHEATHING, 5" THICK, 7" WIDE.
2. SPILLWAY: PLANK SHEATHING, 4" THICK, 8" WIDE.
3. ICE SHIELD: IRON PLATE, 1/4" X 24" X 42", BENT ON
   6" RADIUS AT 32", SECURED WITH IRON SPIKES.
4. SHEET PILING: PLANK SHEATHING, 4" THICK, 8"
   WIDE.
5. SQUARED FIR TIMBER CRIBBING, SECURED WITH
   WROUGHT IRON PINS, 3/4" DIA, 24" TO 36" LONG.
6. 1-3/4" DIA. BORED HOLES SPACED AT APPROX. 40".
   IRON PIPE, 1/8" WALL THICKNESS, WAS INSERTED
   INTO HOLES TO SUPPORT FLASHBOARDS.

PROFILE

0 2 4 6 8 10 FEET

ROCK-FILLED TIMBER CRIB DAM
CONSTRUCTION DETAIL
INTERPRETED FROM REMAINS

EXTANT 4 JULY 1992

SEWALL'S FALLS HYDROELECTRIC
FACILITY - NPS804

FIG. 6.21. A plan view and profile of the rock-filled spillway of the Sewall's Falls Dam.
Courtesy of Dennis E. Howe.

100-foot-long section of the dam to wash out during the night of April
7, 1984, and the Merrimack River has been rushing through the breach
ever since (fig. 6.22).

This massive industrial ruin had originally required about 1,500,000
feet of timber and 20,000 cubic yards of rubble stone to construct (Rivers Engineering 1983), and the abutments and wing walls were made
of squared granite blocks laid in cement. The spillway was constructed
with 12 by 12-inch longitudinal timbers, chiefly hemlock, and 10 by 10-
inch crossties fastened with bolts (Starbuck 1990c: 49, 55). The adjacent
headgate structure had five headgates, opening into a power canal that
carried water for 1,280 feet and then to two brick powerhouses, where
turbines generated power for the city of Concord.

All of the power-generating equipment was removed in the mid-
1960s, and there are no plans to rebuild the dam or return the powerhouses to use. Still, the Sewall's Falls Dam—even with its missing
midsection—continues to be one of the most spectacular industrial survivals in New Hampshire.

The Mount Washington Cog Railway

One of New Hampshire's finest industrial sites is still in operation today:
the Cog Railway has taken passengers up and down Mount Washington
since 1869. Railroads have left many traces across the New Hampshire
landscape—abandoned tracks, trestles, coal loaders, and more—but the
Cog Railway on Mount Washington is a genuine survival of nineteenth-
century technology. "The Cog" is a throwback to earlier times, when
steam power was in use throughout New Hampshire for railroads, boats,
mills, and factories, and today this is the only cog railway that is *entirely*
powered by steam (see Starbuck 1994b, 2003–2004).

Sylvester Marsh of Hampton, New Hampshire, is credited with the
invention of the Cog Railway, and he patented a cog system with big
gears and air brakes (he termed them "atmospheric brakes"). Marsh

formed the Mount Washington Steam Railway Company in 1865, and he constructed the first base station at the foot of Mount Washington in the same year. His track design had two rails positioned four feet eight inches apart and a third "cog rail" between them to accommodate the gears beneath the locomotives (fig. 6.23). The first trial runs began taking passengers for short distances up Mount Washington in 1866, but it wasn't until 1869 that the first train made it all the way to the top.

Early locomotives, such as "Peppersass" (fig. 6.24), had vertical boilers, but beginning in 1874 all subsequent locomotives had four-cylinder horizontal boilers that were slightly raised at the firebox end. They traveled along a railway that has an average grade of 25 percent and a maximum slope of 37 percent. The highest and steepest part of the railway is known as "Jacob's Ladder," where the trestle is inclined at a grade of 37.41 percent and rises to 25 feet above the ground. It takes about an hour for a locomotive to ascend to the top of Mount Washington, pushing a passenger coach with its bumper (fig. 6.25).

For a modern-day industrial archeologist (or anyone else!), a visit to the Cog Railway begins with buying a ticket at Marshfield Base Station,

FIG. 6.23. A close-up of a section of track on the Cog Railway, showing the central "cog rail" between the two running rails, 1994. The cog rail receives the cog gear positioned underneath the locomotive.

FIG. 6.24. "Old Peppersass," fully restored, on display at the Base Station, 1994.

FIG. 6.25. The locomotive named "Base Station" as it pushes a coach on the Cog Railway, 1994.

followed by a noisy, slow, sooty ride in a coach that holds either 48, 56, or 70 passengers. The locomotives shoot great clouds of gray and black smoke into the air, pieces of coal lie alongside the tracks, railroad workers are stained black with soot and grime, and another train departs every hour for the summit. This is one of the greatest industrial archeology experiences in America, and, depending on the weather, it is possible to ride on "the Cog" from May through October of each year.

Some of New Hampshire's older industrial sites have been closed for a long time, or exist only as ruins, while others—such as the Cog Railway and the Swenson Granite Works—continue to be successful, healthy businesses. Industrial archeologists have much to record and preserve in New Hampshire, and there are still many bridges, dams, railroads, mill sites, factories, and quarries that provide excellent research opportunities. New Hampshire's industrial sites include examples of some of America's finest early industries, and these will certainly present a challenge and an opportunity to future generations of scholars.

The high profile given to early industrial structures is key to their continued preservation, as is their adaptive reuse. Many of New Hampshire's finest industrial sites have been placed on the National Register of Historic Places, and Christine Fonda (Rankie) has published a partial inventory of many of them (1994). Some of the sites that are best represented within this listing are historic railroad stations, covered bridges, and (former) textile mills. While many other states have lost their early industrial sites to "progress," New Hampshire has really been quite successful in ensuring the continued survival of its rich industrial base. Through the efforts of state agencies (especially the New Hampshire Division of Historical Resources) and private organizations such as the Northern New England Chapter of the Society for Industrial Archeology there have been many excellent efforts to record and preserve these impressive survivals from our industrial past.

Chapter 7

Marine Archeology in New Hampshire

MARINE ARCHEOLOGY is unquestionably the most exciting sub-field of archeology today, the "new frontier" of archeological research, thanks to its use of cutting-edge technology for locating and recording shipwrecks. Manned and unmanned submarines, remote sensing equipment (especially side-scanning sonar and magnetometers), underwater photography and mapping, and the myriad conservation techniques employed to stabilize finds all add excitement and "science" to a field that initially evolved out of salvage efforts and sport diving. However, it is important to remember that the *best* use of these techniques is to answer questions about maritime history and early seafaring: to understand the construction, outfitting, and navigation of ships, the trade routes they traveled, the cargoes they carried, and the lifestyles of the sailors and passengers who were on board. Marine archeology is most definitely *not* just an opportunity to recover "treasure" or unusually well-preserved artifacts. Rather, it is an extremely difficult field that deals with some very fragile materials, and it requires a great deal of patience to document hull remains with measured drawings (fig. 7.1).

Marine archeology as practiced today was not really feasible until the invention of the aqualung and the demand regulator by Jacques Cousteau and Emile Gagnan during World War II. (The regulator, activated by water pressure, controls the volume of compressed air to the diver based on depth/water pressure.) But no doubt diving on wrecks and salvaging their contents goes back to the very first instances of travel on the water. It also should not be thought that marine archeology deals only with the remains of ships and their cargoes, because techniques of underwater recovery and recordation may be used to better understand *any* type of site that happens to be under the water. Dramatic examples of other categories of underwater sites include the famous Sacred Cenote at Chichen Itza in Yucatan, where Edward Thompson used a dredge to recover Maya gold disks, jade, pottery, balls of resin, and human skeletons from the giant "Cenote of Sacrifice" (Thompson 1932); the buccaneer city of Port Royal, Jamaica, where about 20 acres of houses and streets slid into the water on June 7, 1692, and where salvors and archeologists have been diving ever since (Hamilton and Woodward 1984); and even the remains of bridges and their underpinnings may be documented underwater, such as the log caissons from the 1777 Revolutionary War bridge across Lake Champlain that connected Mount Independence in Vermont with Fort Ticonderoga in New York (Starbuck 1999a: 183–186). The techniques of marine archeology have also been used to record far more mundane sites, including the remains of wharfs and piles of debris or ballast that settled to the ocean floor (fig. 7.2).

FIG. 7.1. Taking measurements in Hart's Cove in New Castle, New Hampshire. Courtesy of David C. Switzer.

FIG. 7.2. Various finds recovered from the seabed survey of Hart's Cove. The artifact distribution pattern seemed to emanate from the Salamander Point area. Courtesy of David C. Switzer.

Underwater investigations offer several very special advantages, because all of the artifacts (and ship remains) form part of the same assemblage; that is, they typically date to just one moment in time. Most of these sites have not been disturbed by man, and many artifacts are complete and well preserved (although ferrous metals often do not survive). And, of course, they provide the best possible evidence for ship construction, shipboard life, and long-distance exchange. The presence of worm damage or pieces of coral may suggest that a vessel once traveled in warmer waters, while a sharp bow suggests that a vessel was built for speed, and ships often reveal many idiosyncrasies that derive from vernacular construction. Still, hulls and artifacts can be badly scattered and crushed when a ship breaks up; the upper parts of the hull and rigging are usually gone; and in warm waters, teredo and limnoria worms typically eat most of the wood. And worst of all, underwater projects are expensive and potentially dangerous! Marine archeology requires plenty of specialized training and experience, and even SCUBA safety courses do not guarantee immunity from the bends, nitrogen narcosis, or entanglement in underwater debris when visibility is poor due to clay or silt in suspension.

The techniques employed in finding underwater sites have changed over the years, and the old-fashioned image of scholars peering through glass-bottomed boats evolved in the 1960s into the use of underwater suction hoses and "telephone booths," and more recently into archeological "prospecting" using side-scan sonar, proton magnetometers, miniature submarines, and remote-control cameras that permit locating and documenting sites on even the very deepest ocean floors. The day will perhaps come when most underwater sites have been located, and in every case the decision will have to be made whether the wreck should be excavated or preserved in situ. The latter has become the preferred option, since modern attitudes toward historic preservation favor leaving every wreck intact; besides, it can be extremely expensive when artifacts or hull fragments are removed from the environment they have adjusted to over many years. Soaking wood artifacts in a solution of polyethylene glycol over a period of years so as to conserve and stabilize them is rarely preferable to simply leaving artifacts underwater in the original spot where they were found.

Marine Archeology in New Hampshire Waters

In many parts of this country, there has been tension between professional marine archeologists and sport divers, and there can be downright hostility between archeologists and salvors. The question that invariably arises is "*Who* should be allowed to explore and to retrieve artifacts from underwater wrecks?" Should archeologists have a monopoly on exploring shipwrecks? While the answer is not always clear, there is no good reason why sport divers and archeologists cannot work together to discover and appreciate the past. New Hampshire state law is quite flexible as it relates to shipwrecks, and the finder gets to hold onto his/her discovery while personally bearing the cost of artifact conservation. Individuals may display in their homes the artifacts that came from underwater wrecks, but they may not sell anything, because all of the artifacts taken from New Hampshire waters legally belong to the people of the State of New Hampshire. It was in 1982 that a law was passed giving the state ownership of all historic resources coming from state-controlled waters.

New Hampshire's very short coastline, just 18 miles, may appear to limit the amount of marine archeology that can be done in the state, but New Hampshire's many inland lakes and rivers are also a possible source of prehistoric and historic sites. In the former category, several prehistoric dugout canoes have been found in New Hampshire lakes, and examples are either on display or in storage at the Libby Museum in Wolfeboro, the New Hampshire Antiquarian Society in Hopkinton, the Museum of New Hampshire History in Concord, and the Division of Historical Resources in Concord (Potter and Switzer 1989).

For the past generation and more, most of the underwater research in New Hampshire has been directed by maritime historian David Switzer of Plymouth State University. Switzer has directed underwater digs in New Hampshire and Maine since 1975 (see the box "Dr. David Switzer, New Hampshire's Distinguished Marine Archeologist"), and together with now-retired State Archaeologist Gary Hume, he assisted in drafting New Hampshire state law as it pertains to underwater sites.

Dave Switzer recently noted that it was in 1981 that Congress passed the Abandoned Shipwreck Act, at which time all states with coastlines were required to prepare plans for how to protect shipwrecks in state waters (Switzer 2003–2004: 93). Switzer was appointed by Gary Hume to be New Hampshire's consulting marine archeologist, and Dave has been on call to advise the New Hampshire Division of Historical Resources ever since. Dave's many projects in New Hampshire really began with a 1980 remote sensing survey of the Isles of Shoals and portions of the Piscataqua River and Great Bay, and this survey even discovered timbers from the 1794 Piscataqua Bridge. Several field seasons were subsequently devoted to work in Hart's Cove in New Castle, which was funded by Sea Grant to Kittery Historical and Naval Museum. This was a joint project involving the Institute for New Hampshire Studies and MAHRI (the Maritime Archeological and Historical Research

★ Dr. David Switzer, New Hampshire's Distinguished Marine Archeologist

David Switzer has been a professor of history at Plymouth State University since 1965, and, though recently retired, he has single-handedly defined the course of underwater research in New Hampshire. Dave received his M.A. and Ph.D. in American history from the University of Connecticut, specializing in maritime history, and he describes his first marine adventure as follows: "While in the U.S. Army as the executive officer of a Nike Anti-Aircraft Missile Battery in Hull, Mass., I convinced the CO et al. that being located on an island near Hull, we should display an antique anchor in front of the headquarters building. Such an anchor, fluke only, was visible at low tide near the island pier. With 55 gal. drums, it took two tides for [the air-filled drums to help] the anchor to break free of the mud. I guess this was my first venture." His next experience was a bit similar because it involved "recovering a dugout canoe from Sebago Lake in the early '60s. Its location was pointed out in waist deep water. I got it out of the sand—8' long, quite narrow. Last I knew it was exhibited at the museum at Univ. S. Maine/Gorham. It had been repaired with a peg in a bored hole; tool marks indicated modern chisels, axes. Theory was it was constructed at a boys' camp ca. early 20th century." (Switzer, personal communication, June 13, 2005).

In 1974 Dave was drawn into the newly created field of marine archeology when he participated as a field school student (at age 40!) on a dig with George Bass and the American Institute for Nautical Archaeology (INA); they were working at the site of a fourth-century A.D. Byzantine (Roman) vessel off the southwest coast of Turkey. Dave gives credit to George Bass for prompting him to pursue a career in marine archeology. Dave

David Switzer in his wet suit while working on the *Defence* project, about 1975–1976. Courtesy of David C. Switzer.

also worked in Scotland in 1975 with Colin Martin, director of the St. Andrews Institute for Maritime Archaeology; this project was at the site of the *HMS Dartmouth*, a small warship sent north to ensure Scottish acquiescence after William and Mary came to the throne of England.

Dave's first opportunity to become the director of his own project came almost instantly, and it was the excavation of the *Defence*, a Revolutionary War, 16-gun privateer that sank in 1779 in an inlet just west of the Penobscot River on the coast of Maine. The *Defence* had been part of the ill-fated Penobscot expedition against the British, when virtually every vessel in the 43-ship American fleet was either scuttled or captured. The wreck of the *Defence* was discovered in 1972 through the use of sonar, and Dave spent six field seasons between 1975 and 1981 excavating the mud-embedded hull. Working together with the INA, the Maine State Museum, and the Maine Maritime Academy, Dave's team discovered evidence for hasty construction "short cuts" and determined that the vessel had a very sharp hull—obviously designed for speed (Switzer 1976, 1981, 1983; Ford and Switzer 1981; Sands 1996: 155–159). The *Defence* was a time capsule in the truest sense, with everything on board dating to the early morning of August 14, 1779, at which time the American crew blew up the brig so that the British would not be able to capture the vessel. After two hundred years, the vessel was still 40 percent intact, and it was the first shipwreck of the American Revolution to be scientifically excavated in situ. More than 40 students and two teams of Earthwatch volunteers assisted Dave in the underwater investigation.

Dave's projects have included the recovery of the bow of the *Snow Squall*, the last surviving American clipper ship, discovered at the end of a jetty in Port Stanley in the Falkland Islands (east of the southernmost tip of South America). The 160-foot *Snow Squall* had been launched from South Portland, Maine, in 1851, and was abandoned in Port Stanley for lack of money for repairs; she was stripped of her sails and rigging in 1864 and converted into a wool warehouse; eventually she became part of a ship graveyard 60 feet from shore (Pave 1985: 65). The *Snow Squall* had been an "extreme" clipper ship, built for speed with a narrow, sharp bow, and this unique survival is an important part of Maine's maritime history. Working with his teammates between 1983 and 1987, Dave helped to document the ship's remains and transported the bow to the Spring Point Museum in South Portland, Maine, where it is now on permanent display.

Dave Switzer has had many years of experience in handling project logistics; conducting historical research; personally doing much of the diving; teaching students and volunteers how to dive safely and to measure and record wrecks underwater; and then disseminating

A youthful Dave Switzer, already preparing to be a marine archeologist while still in high school. Courtesy of David C. Switzer and Peter T. C. Bramhall.

the results through publications and innumerable public lectures. In an archeological specialization that has sometimes been accused of having too many "cowboys," Dave is very much the professional scholar—a historian who has always pursued answers to questions about ship construction and navigation, eschewing the glamour of treasure ships and famous shipwrecks for the knowledge that comes from systematic archeological and historical scholarship.

When recently asked "which has been your favorite project?", Dave's response was,

While all have presented challenges as well as material culture and evidence of the techniques of unknown shipwrights, thereby making possible to add some footnotes re: vessel construction c. 17th–19th centuries, *Defence* has to be high on the list because of its "time capsule" qualities. Also, because of the Rev. War connection, there was a substantial amount of $ support, including National Geographic and Earthwatch, an abundance of field school students, and pretty POSH living conditions at Maine Maritime Academy. In NH the sites at Hart's Cove presented both problems, finds that provided general dates, and interesting structural features. However, identity remained a mystery, and then there was *Lizzie Carr*, about which we knew just about everything. A nice sort of closure. (Switzer, personal communication, June 13, 2005).

FIG. 7.3. Scene of Hart's Cove with Salamander Point in the background. The site of the shallop-type vessel is below the moored boats in the center of the picture. Courtesy of David C. Switzer.

FIG. 7.4. A field school student diver working at Salamander Point. Courtesy of David C. Switzer.

Institute). In Hart's Cove Dave recorded a shallop-type vessel and then investigated the site of a late-seventeenth- or early-eighteenth-century merchant-type vessel at Salamander Point (figs. 7.3–7.5). Dave's other New Hampshire projects have included a 1984 survey of the steamboat *Stella Marion* in Newfound Lake; a 2001 underwater study of the Enfield Shaker Bridge; and the investigation of the schooner *Lizzie Carr* in Hart's Cove between 1998 and 2001.

FIG. 7.5. A Rhenish stoneware chamber pot recovered from the Salamander Point site.

The Hart's Cove Wreck

One of the most common types of early colonial boats in New England was the shallop, a small craft with one or two masts that was double-ended and used as a fishing craft until it was replaced by derivative boat types. But frequency of use in the past does not always mean that we have a detailed knowledge of construction methods, and little information is available today about how shallops were put together. It was thus fortuitous when Switzer's 1980 survey located an anomaly in the mouth of the Piscataqua River at a depth of 35 feet. The end of a keel, some frame ends (fig. 7.6), and some scattered seventeenth- and eighteenth-century artifacts all provided evidence for a buried vessel. As archeological testing began in 1982 and 1983, Switzer determined that it was a 30-foot-long craft with a round bottom and a smooth-sided hull (Switzer 1985, 1991, 1994). The excavation of the wreck took place between 1985 and 1987, and again in 1991, followed by a seabed survey in 1992. Switzer and his team completely exposed and drew the remains of the possible shallop (fig. 7.7), discovering a highly irregular vessel with lots of evidence for vernacular construction.

The Hart's Cove wreck showed plenty of signs of looting by treasure-hunters, and the large number of artifacts scattered across the remains of the hull—possibly from later dumping—made it difficult to determine which finds were actually associated with the wreck. There were many sherds of European and English stoneware, North Devon sgraffito and gravel-tempered ware, tobacco pipe bowls and stems, "onion" bottle fragments, part of an Iberian-type storage jar (eighteenth century), and even

FIG. 7.6. The shallop site with frames or ribs exhibiting a curve; therefore, the remains are not those of a gundalow (the sides of which would be vertical, with no curvature). Courtesy of Gary Carbonneau, Underwater Photography.

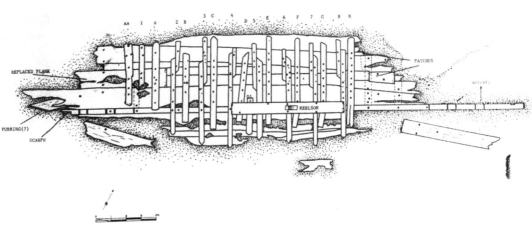

FIG. 7.7. Measured drawing of the remains of the shallop-type vessel. The bow is toward the left; the extant structure comprises the starboard (right) side. Prepared by Sheli O. Smith.

a fragment of Westerwald mug bearing a medallion with the inscription "REX WILH . . . " (referring to William III of Orange, 1689–1704).

On a lighter note, every year that Dave Switzer and his crew returned, they would find the same striped bass swimming in Hart's Cove, and they enjoyed feeding their "friend" (Switzer, personal communication, January 24, 2005). One day a fisherman stopped by and asked the archeologists if they had seen any striped bass. Dave replied, "There aren't any *here*, but if you go around the corner, there are *lots* of them." The fisherman left, and the marine archeologists were delighted to have saved their pet fish!

So was this the first example of a shallop ever discovered archeologically in the Americas? Switzer has estimated that the total length of the vessel in Hart's Cove was between 35 and 38 feet, with a maximum beam of eight feet and a hull depth of about four feet. Based on dimensions, hull characteristics, and known historical references to shallops, this New Hampshire wreck really does appear to have been a shallop. It is difficult to date the vessel precisely, but the bores of the English tobacco pipes, when measured, provide a mean date of A.D. 1693, so it appears that the Hart's Cove wreck does fall in the late seventeenth century. While incomplete and somewhat enigmatic, this wreck is nevertheless an extremely rare example of what was most likely one of the most common types of vessel in early New England.

The Steamboat *Stella Marion*

In 1984 Switzer, working through the Institute for New Hampshire Studies (INHS) at Plymouth State, received a permit to survey the wreck of a steamboat, the *Stella Marion*, in Newfound Lake (Switzer 1985). The 50-foot *Stella Marion* was built in 1900, sank at her mooring during a fire in 1915, and was essentially untouched until her discovery by a sport diver. While perhaps not a glamorous type of site, the *Stella Marion* (named after the two daughters of her owner, Ambrose Adams) represents a lake steamer that delivered mail, towed barges, and so on during her short 15-year life (fig. 7.8). The INHS offered a field school in nautical archeology at the site in the summer of 1985 (fig. 7.9).

Switzer's survey efforts discovered that much of the wreck had survived, and he prepared detailed drawings of the hull and its equipment (figs. 7.10–7.12). No excavation was conducted, but the engine and the boiler were still intact inside the hull (fig. 7.13). It would have been costly to retrieve and conserve the two-cylinder, high-pressure engine, but this is a good example of the type of vessels that most probably lie sunken within several of New Hampshire's lakes.

FIG. 7.8. The *Stella Marion*, probably sunk between 1910 and 1916. The remains of the searchlight mounted in the wheel house were recovered from the bow area. Courtesy of David C. Switzer.

FIG. 7.9. Plymouth State field school students collecting measurements at the *Stella Marion* site. Courtesy of David C. Switzer.

FIG. 7.10. Site plan of the *Stella Marion* site. Prepared by David C. Switzer.

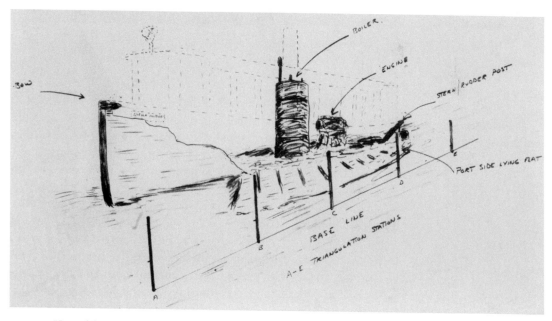

FIG. 7.11. View of the *Stella Marion* site. The vertical poles are triangulation stations used to produce a measured plan of the site (fig. 7.10). Prepared by David C. Switzer.

FIG. 7.12. Base of an oil lamp recovered from the *Stella Marion*. According to Dave Switzer, it was perhaps the culprit that caused the fire that sank the steamer in about 1916. Courtesy of David C. Switzer.

FIG. 7.13. Boiler and steam engine found inside the hull of the *Stella Marion*. Courtesy of David C. Switzer.

Dave's only regret afterward was that the *Stella Marion* "might have had an experimental archeological conclusion—a replica constructed from structural and dimensional data recorded and powered by her rejuvenated steam engine. Too much $ involved and not enough interest at the time" (Switzer, personal communication, June 13, 2005).

The Enfield Shaker Bridge

David Switzer's survey work under the auspices of the Institute for New Hampshire Studies has located scattered timbers, popularly known as "Jack Straw," from the 2,600-foot-long Piscataqua Bridge crossing Great Bay from Newington to Durham (which was in operation from 1794 to 1855), but there are other New Hampshire bridges that have left their own underwater record. One of the most interesting of these is the Shaker bridge constructed between 1847 and 1849 in Enfield, New Hampshire. Many photographs and postcards depict the famous Shaker Bridge across Lake Mascoma, and when the Hurricane of 1938 carried away the roadway portion of the bridge, it was only the cribwork that remained intact and underwater. (The timber cribwork was built on the ice during the winter of 1848 and sank to the lake bed when ice melted in the spring.) Unfortunately, when construction occurred in 1939 to allow replacement bridge footings to go in, much of the remaining cribbing was destroyed.

It was not until 2001 that David Switzer and Brendan Foley had the opportunity to dive on the remains of the bridge and to document for the New Hampshire Department of Transportation—with a mosaic of underwater photographs—what had actually survived of the bridge (fig. 7.14). Their photomosaic and mapping have revealed that only part

FIG. 7.14. The Enfield Shaker Bridge Project. David Switzer prepares a list of tasks for field school diver Bert Sutcliffe. Courtesy of Katherine Donahue.

of the western face of the cribwork still rests in Lake Mascoma—the dredging conducted in 1939 destroyed all other vestiges of the Shaker Bridge. In one way their survey was rather typical of underwater archeology in New Hampshire lakes, because their visibility was only two to four feet, making it harder to take photographs and necessary to be careful while working around underwater debris.

Underwater work on the Enfield Shaker Bridge revealed that the construction of the 1939 bridge caused logs and boulders from the Shaker Bridge's cribwork to be scattered across a large area of the lake bed. One of these surviving logs, which was notched to receive other logs, measured 20 inches in diameter and nine feet long. The one intact, in situ section of cribwork from the Shaker Bridge is about 30 feet long; "the height is fourteen feet from lake bottom. The logs diminish in diameter as the cribwork ascends vertically" (Switzer and Foley 2001: 2). It is unfortunate that so little remains of the Shaker Bridge, but Switzer and Foley have carefully documented the traces that are left.

The Schooner *Lizzie Carr*

The most recent marine project in Dave Switzer's long career involved the hull of a ship found on Wallis Sands Beach in Rye. A storm in 1998 exposed timbers from a 60-foot-long hull lying buried on the beach, and Switzer held his first field season there in 1999. A local beach resident, Glenne Ford, soon presented him with a photograph that identified this wreck as "Schooner Lizzie J. Carr, Jan. 12, 1905." Historical research then identified the *Lizzie Carr* as a two-masted schooner built in Maine in 1868. The schooner was chiefly involved in the domestic coastwise trade, and when she was driven ashore in a storm in 1905, she had been carrying a cargo of 270,000 feet of lumber that was left scattered all over the beach.

With the identity of the wreck no longer a mystery, fieldwork was carried out in 2000 and 2001 as Switzer and his students removed sandy overburden, measured and drew, and then removed a section from the hull for transport to the New Hampshire Seacoast Science Center for conservation and display. While more of a "beach" than an "underwater" project, this site was challenging because it was underwater at high tide. This prompted Switzer and his team to bring in sand bags to build a berm on two sides of the wreck, and fresh sand had to be removed every day during work hours. Oak trunnels (tree nails) had fastened the yellow pine planks to oak frames (ribs) (Switzer 2003–2004: 99). Three bronze spikes were also found, which probably held a plank in place while holes were being drilled for the trunnels.

FIG. 7.15. Preparing the hull of the *Lizzie Carr* for its "big lift" to the Seacoast Science Center. Courtesy of David C. Switzer.

FIG. 7.16. A section of the *Lizzie Carr*'s hull on display in the Seacoast Science Center. Courtesy of David C. Switzer.

The *Lizzie Carr*, at 286 tons, was 138 feet long, and as a carrier of cargo in the coastal trade, she was an important representative of the time when schooners ruled the waves and made maritime commerce successful. Fortunately for Switzer's team, the Seacoast Science Center at Odiorne Park was undergoing renovations and agreed to accept a portion of the *Lizzie Carr*'s hull as a major new exhibit. The hull section that was to be moved was much heavier than expected—about two tons—but a tractor with a backhoe lifted the hull from the beach (fig. 7.15) and a flatbed truck moved the section to the Science Center (fig. 7.16). After treatment with the fungicide Boracare, plus a small amount of the preservative polyethylene glycol, the *Lizzie Carr* exhibit opened in May of 2004. As Switzer has noted, the *Lizzie Carr* is "a legacy of the last days of American sail" (2003–2004) and a very proud part of New England's maritime heritage.

Only a handful of archeologists have been trained in the proper techniques for locating, drawing, and excavating underwater sites, and New Hampshire is extremely fortunate to have the services of David Switzer, one of the leading scholars in this exciting field. As Switzer's writings often suggest, much more mapping of the seabed still needs to be carried out, workshops are needed to train sport divers and archeology students in survey techniques, and underwater preserves maintained by the State of New Hampshire would help to make these discoveries more accessible to the diving public. With systematic help from the sport diving community, it should be possible to locate more wreck sites and to increase public awareness of the need to preserve the remains of shipwrecks.

Switzer's 1980 remote sensing survey of the Piscataqua basin is still the only systematic regional survey ever conducted in New Hampshire, and no doubt there are other wrecks on the coast of New Hampshire and in inland lakes that warrant research. Still, the need for substantial funding for sophisticated equipment and for the conservation of waterlogged materials has always limited the number of underwater projects that may be conducted. This is perhaps for the best, but the search for New Hampshire shipwrecks will surely continue, and I am certain that it will be David Switzer leading the way!

Epilogue

So, You Want to Be an Archeologist?

T**HE POPULAR IMAGE** of an archeologist has definitely changed. When I was a student, archeologists were supposed to be bearded, balding, pipe-smoking, tweed-wearing scholars, and prolific consumers of sherry or scotch. This of course also implied they were supposed to be men! In time, with some assistance from Hollywood and Stephen Spielberg, this image livened up a bit, and the next archeological stereotype was that we would wear Indiana Jones–style felt hats (nobody actually wears those hats—they're too hot!), drive Jeeps or Land Rovers (when we're not riding horses), be rather scruffy-looking (yet tremendously appealing to women), still drink too much (some parts of the image don't change!), and be instantly ready to go anywhere for a good adventure. I don't think I know any archeologists who fit that description exactly, but I *do* know lots of archeologists who have delighted in being compared to Indiana Jones. In fact, I understand from my women colleagues that they're quite amused (and clearly delighted) when *they* are compared to Indiana Jones (fig. Epi.1)!

What is it like to be an archeologist today? Even though I miss the glamour of old-style archeology, I have to acknowledge that over the past generation or two, archeology has finally grown up. What was once a fun hobby for those with leisure time has become a serious academic discipline, and thousands of men and women are now pursuing careers in archeology. But employment opportunities have changed, and the university-based or museum-based archeologist who teaches, prepares exhibits, and dreams all winter of "the dig next summer" is now in the minority. We have been joined by a host of archeologists in regulatory positions who are charged with protecting our past—which means they push a lot of paper!—and by thousands of those who conduct cultural resource management surveys. Archeology has entered the business world. Many of us do not have a lot of job security, but that's not why you become an archeologist. Most of us are positively passionate about what we do, and most never want to retire. The chance to learn something new, to "connect" with someone who lived hundreds or thousands of years ago, is a very special thrill that we cannot live without.

Is it possible to define what a "good" archeologist is? As I teach my classes of students, I have always stressed one thing: you must write. A good archeologist writes up everything. When we excavate and remove pieces of the past, when we dig archeological sites and thus destroy what is in the ground, it places us under a serious ethical obligation to publish the raw data and to interpret what we have found.

FIG. EPI.1. The "face" of modern archeology—Justine "Brownie" Gengras (left), Victoria Bunker (waving), and Millee Bolt (seated) at Garvin's Falls.

FIG. EPI.2. White Mountain National Forest archeologists and paraprofessionals excavate, record, and map shovel test pits in advance of a construction project in Campton, New Hampshire. Courtesy of the White Mountain National Forest.

I once knew the dean from one of New Hampshire's colleges quite well, and he described to me how he had given a sabbatical to the one archeologist on his faculty for the purpose of getting some of his old site reports written up. Instead of writing, the archeologist went out and dug another site in the town where he lived, thus adding to his (unpublished) backlog. Musing on the sabbatical, my friend the dean commented, "Never again." Any archeologist who has more than one or two unpublished sites should be banned from further digging until the results are written up! Archeologists *do* have that much of an obligation to the past.

Archeology in New Hampshire Today

As this book points out, New Hampshire has met the challenges facing modern archeology in positive ways, blending "pure" research with cultural resource management surveys and volunteer opportunities in New Hampshire's SCRAP program and the White Mountain National Forest (fig. Epi.2). New Hampshire's archeologists have discovered exciting prehistoric, historic, industrial, and marine sites, and the New Hampshire Archeological Society (NHAS) and the Division of Historical Resources (DHR) have traditionally led the way in guiding field and laboratory work in the state. Since the formation of the NHAS in 1947, archeology in New Hampshire has come a long way, and hundreds of site locations have been listed in the site files of the DHR, hundreds of sites have been published upon or described in technical reports, and thousands of students and volunteers have learned to more fully appreciate the past by participating on digs. There is room for everyone to be engaged in the field of archeology, at least in a volunteer capacity.

Is there any site in New Hampshire that I personally would still like to dig? Yes, there is: the site of my first ancestor in America, Edward Starbuck, who settled with his family on Dover Point in 1635. If the cellar hole still exists in someone's backyard, and if it wasn't destroyed by the construction of the Spaulding Turnpike, then I would absolutely love to do an excavation there! But my point is that archeology has the potential to be meaningful to us on a personal level. I do not believe that archeology would be exciting to so many of us if it were not somehow relevant to our own lives. Through archeology, we draw closer to those who were here before us; we can better appreciate earlier styles of living; and we can better understand our own place in the long-term procession of human cultures. We don't necessarily have to dig up our very own ancestors (although that would please me no end!), but the "things" of the past really do speak to us.

In contemplating the archeology that has been practiced in New Hampshire, my one lingering regret is that there are quite a few sites that have never seen a thorough publication of their final results (for example, the Hunter site and Weirs Beach, among others). This is certainly a warning to the rest of us that we should not engage in digging unless we are prepared to see a project through to completion—and that means complete publication of all results in a timely fashion. As New Hampshire archeology further matures, we need to spend increasing amounts of time analyzing and reanalyzing "old" collections rather than always digging new sites. The preservation ethic has become central to all archeological work in the United States, and we should gradually reduce the amount of digging we do and only dig when there are significant new research questions that need answering (or when a site faces destruction). Removing an archeological site can never be justified if it is just to "get students out of the classroom" or to earn an instructor some extra summertime income!

Future Directions: We Need to Preserve Our Past!

Archeological sites are being destroyed at an alarming rate all over the world (Meyer 1973, Vitelli 1996), and even in New Hampshire many sites no longer exist due to bridge or road construction, housing developments, stream erosion, or treasure hunting. Sometimes it may not be realistic to try to avoid *every* archeological site, but we need a well-informed population that will report site locations to appropriate state agencies and that will not tolerate developers placing housing developments on top of ancient sites. Until everyone "thinks like an archeologist," sites will continue to be lost.

The current spate of archeology programs on cable television is but one of the indicators that the public at large is absolutely fascinated with archeology, and this should enable us to enlist more public support: we need to achieve a better balance between the needs of development and those of historic preservation. We archeologists often say to each other that in another generation or so, all of the "good" archeological sites will be gone, and our successors will only be doing collections research. I do not believe that day will ever actually come. Rather, I really believe that the public's support for archeology will enable us to keep digging and to keep learning, even as we find ways to better protect archeological sites so that they may be appreciated and studied by future generations.

One way to achieve this is by joining and promoting the goals of the Archaeological Conservancy headquartered in Albuquerque, New Mexico, the only national nonprofit organization whose mission it is to acquire archeological sites and thereby protect them from development. Thus far, they have preserved more than three hundred sites in the United States, including one site in New Hampshire, and the potential exists for them to buy many more. Financial contributions made to the conservancy are a great way to help protect our past, especially if you are unable to join us in the pits!

Appendixes

Appendix 1

Where to Go for Further Information

It is probably true that almost every child grows up wanting to become an archeologist (or else wanting to dig for dinosaur bones!). Typically, it is not until it is time to apply for college that career goals change into something "more practical." However, whether you are one of those still planning to become an archeologist or—more likely—wishing you had, there are many opportunities to become more involved in this exciting field.

Here are answers to some of the more commonly-asked questions:

What branch of New Hampshire state government deals with archeology?

New Hampshire Division of Historical Resources
19 Pillsbury Street, 2nd Floor
Concord, NH 03301-2043
(603) 271-3483 or 271-3558
http://www.nh.gov/nhdhr

If I find an archeological site (or artifacts) and would like to report it, whom should I see?

The New Hampshire State Archaeologist is Dr. Richard A. Boisvert (he is based in the Division of Historical Resources). His duties are many, but if you know of a site that is in danger from erosion or treasure hunting, he is the first one to call. Dr. Boisvert may be contacted by email at Richard.Boisvert@dcr.nh.gov or at the DHR.

You may also want to contact the archeologist(s) at one of the following colleges and universities, all of which offer archeology courses:

Department of Anthropology
408 Silsby Hall, HB 6047
Dartmouth College
Hanover, NH 03755-3547
(603) 646-3256
http://www.dartmouth.edu/~anthro/

Department of Anthropology
Crestview Bldg., College Rd., PO Box 60
Franklin Pierce College

Rindge, NH 03461
(603) 899-4260
http://www.fpc.edu/pages/Academics/behave/anthro/anthro.html

Anthropology/Sociology Program
Social Science Department
Plymouth State University
17 High Street, MSC #39
Plymouth, NH 03264-1595
(603) 535-2335
http://www.plymouth.edu

Department of Anthropology
Huddleston Hall, 73 Main Street
University of New Hampshire
Durham, NH 03824-3552
(603) 862-1864
http://www.unh.edu/anthropology

Or, if you have discovered a site in the White Mountain National Forest, you may wish to contact their staff archeologist:
Karl Roenke
Heritage Resource Program Leader
White Mountain National Forest
719 Main Street
Laconia, NH 03246
(603) 528-8773

I would like to take part on an archeology dig in New Hampshire. What should I do?
Some of the colleges and universities listed above offer summer archeology field schools for credit, and usually volunteers are accepted as well. Also, the Forest Service conducts summer digs in the White Mountains.

Another option is to take part in New Hampshire's State Conservation and Rescue Archaeology Program (SCRAP). Many of the projects described in this book were sponsored by SCRAP, which is "a public participation program for archeological research, management, and education." Field schools (college credit available) are offered every summer at sites in New Hampshire, and laboratory training and volunteer opportunities are available throughout the year. Contact SCRAP at the Division of Historical Resources, 19 Pillsbury Street, Box 2043, Concord, NH 03301–2043. The SCRAP website is http://www.nhscrap .org, and you may send emails to nhscrap@nhdhr.state.nh.us.

I would like to hear talks about archeology and meet other people who love archeology. Are there any local archeology groups I can join?

New Hampshire Archeological Society, Inc.
Attn: Treasurer
PO Box 406
Concord, NH 03320-0406
http://www.nhas.org/
Membership: Individual ($20); Institutional ($30); Senior or Student ($18); Lifetime ($425)

New England Antiquities Research Association
NEARA Membership, Betty Peterson
1199 Main Street
Worcester, MA 01603
http://www.neara.org/
Membership: Active ($30); Patron ($40); Sponsor ($60)

Northern New England Chapter of the Society for Industrial Archeology
Carolyn Weatherwax, Treasurer
305 Heritage Way
Gansevoort, NY 12831
http://www.siahq.org/chapters/chapt.html
Membership: Regular ($10); Student ($5); Lifetime ($100)

My family would like to visit a local museum that has archeology exhibits. Where can we go?

Museum of New Hampshire History
The Hamel Center
6 Eagle Square
Concord, NH 03301-4923
(603) 228-6688
http://www.nhhistory.org/museum.html

Strawbery Banke Museum
PO Box 300
Portsmouth, NH 03802
(603) 433-1100
http://www.strawberybanke.org/info/information.html

Mount Kearsarge Indian Museum
Kearsarge Mountain Road

PO Box 142
Warner, NH 03278-0142
(603) 456-3244
http://www.indianmuseum.org

Manchester Historic Association
Millyard Museum
Old Amoskeag Mill #3
200 Bedford South
Manchester, NH 03101
(603) 622-7531
http://www.manchesterhistoric.org

Anthropology Museum
Phillips Exeter Academy
Academy Building
Front Street
Exeter, NH 03833-2460
(603) 777-3452
Open during school hours during the academic year or by
appointment. You may email Donald Foster at dfoster@exeter.edu
to arrange for a visit.

Appendix 2

Locations of Archeological Collections in New Hampshire

By Patricia W. Hume and Donald Foster (1994)
Revised by Patricia W. Hume (2005)

This appendix describes the archeological materials held by the New Hampshire Archeological Society (NHAS), as well as the content and location of numerous archeological collections found throughout the state. It is presented in four parts so as to provide a variety of access routes to these data.

NHAS archeological materials are housed at Phillips Exeter Academy (PEA). These consist of donated collections as well as site materials from NHAS-sponsored excavations. In addition, the society maintains files on all known New Hampshire archeological sites and collections.

The numerous archeological collections found throughout the state consist of material in private hands as well as New Hampshire collections donated to private and public institutions. The information presented in this part of the appendix is based on a mail survey of New Hampshire schools and historical societies, libraries, and museums, as well as data available in the NHAS files housed at PEA. This listing is by no means complete.

Artifacts from contract archeology projects, as well as those from SCRAP summer excavations (fig. App.2.1), are housed at the SCRAP Laboratory in Concord, NH (administered by the New Hampshire Division of Historical Resources).

Appendix 2 is divided into four parts. These are entitled:
1. New Hampshire Collections Housed and/or Listed in the Files at Phillips Exeter Academy
2. Provenience of Artifacts by Towns
3. Sites Described in *The New Hampshire Archeologist* since 1980
4. New Hampshire Collections Housed at Private and Public Institutions

Part 1 lists the New Hampshire archeological materials housed at PEA by site and by collection name, as well as sites and collections

FIG. APP.2.1. Patricia Hume, Chair of the NHAS Site Files Committee, taking notes at the Sewall's Falls excavation in Concord.

found in the NHAS files. A code is included that describes the artifact content and availability of slides. Part 1 also includes information on out-of-state collections and files.

Part 2 is an alphabetical list of towns where the artifacts were originally found. The current location of the collections containing these artifacts is indicated.

Part 3 is an alphabetical list of towns with sites reported in *The New Hampshire Archeologist*, the bulletin of the NHAS. A code is included that describes the abbreviated entries.

Part 4 lists the schools, historical societies, libraries, museums, and other institutions that currently house archeological collections.

Part 1. New Hampshire Collections Housed and/or Listed in Files at Phillips Exeter Academy

Housed at PEA

SITES

Adam's Point: (NH40-41) (PEA, UNH, DHR) (Durham) (artifacts: PP, PO, ET, SC, etc.)

Cote, Al: (NH47-33) (PEA) (Exeter) (artifacts: some)

Dennis Farm: (NH17-8) (PEA and V. Pushee) (Lyme) (NHAS bull., vol. 32, 1991) (artifacts: numerous PP, SC, ET, PO, etc.)

Eddy: (NH38-6) (DHR) (Manchester) (SL-AV-PEA) (NHAS newsletter, vol. 1, no. 2, fall 1985) (artifacts: numerous PP, FL, PO, ET, etc.)

Edgerley's: (NH47-6) (PEA) (Hampton Falls) (artifacts: PP)

Fort Rock: (NH47-9) (PEA) (Exeter) (artifacts: some)

Garvin's Falls: (NH38-1) (PEA and Plymouth State) (Concord) (three
   articles: NHAS bull., vol. 26, 1985) (artifacts: numerous PP, PO, FL,
   ET, etc.)

Mine Falls: (NH44-10) (PEA) (Nashua) (artifacts: several)

Oaklands: (NH47-8) (PEA) (Exeter) (artifacts: several)

Pickpocket Falls: (NH46-2) (PEA) (Exeter) (artifacts: some)
   (NHAS-SL-FI)

Powder House Point: (NH47-60) (PEA) (Exeter) (artifacts: several)

Rock Shelter: (No no.) (PEA) (Exeter) (artifacts: some)

South Hampton: (No no.) (PEA) (near Aspen Hill, South Hampton)
   (artifacts: some) (NHAS-SL-FI)

Sportsman's Pond: (NH49-1) (PEA) (Fitzwilliam) (artifacts: some)

Stanley: (NH47-18) (PEA) (Exeter) (NHAS bull., vol. 23, 1982) (artifacts: some)

Stetson Farm: (NH40-1) (PEA) (Newfields) (artifacts: some)

Stevens Brook: (NH11-12) (PEA) (Shelburne) (artifacts: ET, FL)

Wadleigh Farm: (NH47-59) (PEA) (Exeter) (artifacts: some)

White: (NH47-103) (PEA) (Exeter) (artifacts: PP, etc.)

Wiswall Falls: (NH40-10) (PEA) (Newmarket) (artifacts: ST, etc.)

Yeaton, Maurice, Farm: (NH38-60) (PEA) (Epsom) (NHAS bull., vol.
   32, 1991) (artifacts: 1 PP, 7 SC, 2 HS, 1 TR, 2 CO, 136 FL, 687 PO,
   and hist. clay BE)

PERSONAL AND SURVEY COLLECTIONS

Belanger: (PEA) (Manchester) (artifacts: 37 FL, 24 PO)

Chesley, Dennis: (PEA) (Milton) (2 sites on Northeast Pd.) (artifacts:
   16 PO, several FLs, bone)

Cohas Brook Survey: (PEA) (Manchester) (artifacts: several)

Colby, Solon B.: (on loan at PEA) (Merrimack Valley) (artifacts:
   numerous, including almost all time periods: many PP, PO, ST, etc.)

Collins, Betty: (PEA) (Manchester, Amoskeag area) (artifacts: 34 PO,
   34 FLs)

Crosbie, Laurence: (PEA) (New Hampshire and OOS) (artifacts: several PP, DR)

Davis, Paul: (PEA) (Rye) (Site NH40-2) (artifacts: 16 PO, 3 ST, 25 FL,
   CT, AS, CO)

Demar, John: (PEA) (Laconia, Weirs Beach area) (artifacts: 20 POs,
   PPs)

Derry Survey: (PEA) (Derry) (artifacts: some)

Finch, Eugene, and Davis Finch: (PEA) (artifacts: numerous, from
   several sites)

Hardy, David: (PEA) (Kingston, Lyndeboro, Manchester, Ossipee) (artifacts: Kingston: PO, 20 FL, 3 PP, CO; Lyndeboro: glass; Manchester: FL; Ossipee: 164 FL, 2 PP, CO)

Hatch: (PEA) (donated to NHAS—unknown origin) (artifacts: 1 PE, 1 AX, 1 PQ)

Holmes, P.: (PEA, personal, and Plaistow Hist. Soc.) (Kingston, Manchester, Newton Junction, Plaistow) (NHAS bull., vol. 23, 1982) (artifacts: numerous, from several sites, including PP, PO, FL, ET, etc.)

Hurd: (PEA) (?) (artifacts: 2 AX, PP)

Johanson, Sarah: (PEA) (Manchester, Amoskeag area) (artifacts: 16 FLs, 1 SC)

Lisbon: (PEA) (Northfield, Pittsfield, Plaistow, Tilton) (artifacts: 2 boxes)

Magnuson: (PEA) (Berlin, Manchester) (artifacts: 2 boxes)

Michaud, Allen: (PEA) (artifacts: numerous, from several sites)

Prindle: (PEA) (Clark's Island, Lochmere) (artifacts: several) (SL-AV-PEA) (NHAS-SL-FI)

Rhodes, K.: (PEA) (donated to NHAS 1993) (Manchester) (artifacts: Smyth Site: surface coll.)

White, William: (PEA) (Concord, Epping, Exeter, Greenland, Hampton, Hampton Falls, Kensington, Kingston, Lee, Litchfield, Newfields, Nottingham, Raymond, Seabrook) (artifacts: numerous PP, ST, etc.) (NHAS-SL-FI)

Williams, J.: (PEA) (Exeter) (artifacts: some) (NHAS-SL-FI)

Listed in NHAS Collections Files

SITES

Argillite Workshop: (NH20-6) (UNH) (Tamworth) (NHAS bull., vol. 32, 1991) (artifacts: FL, etc.)

Broad Brook: (NH41-17) (Colony House Museum, Keene) (Winchester) (NHAS bull., vol. 25, 1984) (artifacts: hist.)

Campbell: (NH45-73) (DHR) (Litchfield) (NHAS bull., vol. 29, 1988) (artifacts: PP, PO, SC, ST, etc.)

Canterbury Shaker Village: (on site) (Canterbury) (NHAS bull., vol. 31, 1990; vol. 37, 1997; vol. 43/44, 2003–2004) (artifacts: hist.)

Dodge, Jim: (NH37-12) (personal) (New Boston) (NHAS bull., vol. 28, 1987) (artifacts: 161 PO, 33 FL, 1 PR, 1 PP, CO)

Drake: (NH31-20-2) (DHR) (Belmont) (artifacts: numerous; PP, PO, ST, etc.)

First Fort: (NH31-34) (PSU) (Boscawen) (NHAS bull., vol. 26, 1985) (artifacts: hist.)

Garvin's Falls: (NH38-1) (PEA and PSU) (Concord) (NHAS bull., 3
articles, vol. 26, 1985) (artifacts: numerous; PO, FL, etc.)
Hunter: (NH28-3) (H. Sargent) (Claremont) (artifacts: some)
(SL-AV-PEA)
Meadow Pond: (NH49-1) (?) (Fitzwilliam) (artifacts: some)
Musquash Pond: (NH52-5) (Pat Hume) (Hudson) (artifacts: 1 PP,
3 FL, and hist.)
Mystery Hill: (NH46-1) (on site) (Salem) (artifacts: PP, PO, FL, ST,
CA, and hist.)
New England Glassworks: (NH43-5) (PSU) (Temple) (NHAS bull.,
vol. 27, 1986) (artifacts: hist.)
Rocks Road: (NH47-21) (UNH) (Seabrook) (NHAS-SL-FI) (NHAS
bull., vol. 28, 1987) (artifacts: several)
Rodonis: (NH45-6) (DHR) (Litchfield) (NHAS bull., vol. 29, 1988)
(artifacts: PO, SC, FL)
Russell's Inn: (NH29-1) (Sunapee Hist. Soc.) (Georges Mills) (artifacts:
numerous, many time periods)
Sewall's Falls Sites: (NH31-30, 31-40, 31-43) (PSU) (Concord)
(NHAS bull., vol. 23, 1982; vol. 25, 1984; vol. 26, 1985; vol. 29,
1988; vol. 38, 1998) (artifacts: numerous; PP, PO, etc.)
Smith Farm: (NH46-17) (UNH) (Brentwood) (NHAS bull., vol. 22,
1981) (artifacts: some)
Smolt: (NH45-67) (DHR) (Litchfield) (NHAS-SL-FI) (NHAS bull.,
vol. 24, 1983) (artifacts: SC, GR, FL)
Wadleigh Falls: (NH39-1) (UNH) (Lee) (NHAS bull., vol. 22, 1981)
(artifacts: numerous)
Wasserman Park Sites: (27-HB-65 thru 69) (Dickinson and DHR)
(Merrimack) (artifacts: PP, PO, ET, etc.)
Wentworth, Governor: (NH27-2) (PSU) (Wolfeboro) (NHAS bull.,
vol. 30, 1989) (artifacts: hist.)
Wiggin: (NH40-43) (DHR and P. Wiggin) (Stratham) (artifacts: pre-
hist. and hist.)

Personal, Institutional, and Survey Collections
not Housed at Phillips Exeter Academy
Adams, Lincoln: (Boscawen Historical Society and some in Vt.
and Fla. with sons) (Boscawen) (artifacts: 10 PP, 2 PI, 9 BL, etc.)
(SL-AV-PEA)
Antiquarian Society of the Mount Caesar Union Library:
(ASOTMCUL) (Swanzey) (artifacts: 38 PP or SC, 1 PL, 2 GO,
2 AD, 2 PE, AT, HO, 2 AX, 8 GF)
Barlow, Raymond: (personal) (Windham area) (artifacts: 16 PP, 1 PI,
1 GT, 1 BE, 1 AD, 1 PE)

Batchelder: (personal) (Kensington) (artifacts: some)

Beaudoin, Gary: (personal) (Manchester) (artifacts: KN, several PP, some PO)

Berlin City Library: (BCL) (Berlin)

Berry, Clyde: (Manchester Historical Assoc.) (collected in several towns) (PEA has OOS) (artifacts: numerous, from several sites) (separate collection from out of state donated 1992 to NHAS and housed at PEA; see OOS list)

Blaisdell, Maurice: (personal) (Pembroke) (artifacts: 23 PP, 1 PR, 3 PO)

Boscawen Historical Society: (BHS) (Boscawen) (artifacts: L. Adams and C. Folsom Collections)

Boyd, Robert: (personal) (Plymouth area) (artifacts: 1 PE, 1 AX) (NHAS-SL-FI)

Brown, Winifred: (personal) (Derry) (artifacts: 1 AX, 1 PE, 1 WS, 1 PP)

Canaan Historical Society: (CHS) (Canaan) (artifacts: 1 PP, 1 DR, 1 SC, 1 HO, 1 fish spear P)

Chapin, Levi: (?) (North Walpole area) (artifacts: 2 FE, 1 MA, 1 AD, 1 AX, 4 PP, 4 BL, 2 PI)

Cheswell, Wentworth (1746–1817): (?) (NH) (artifacts: STs)

Colby, Perley: (MHA) (Litchfield) (artifacts: numerous)

Colony House Museum: (CHM, Keene) (artifacts from Broad Brook Site in Winchester)

Conway Historical Society: (CHS) (Conway area)

Corliss, Less: (personal) (Concord) (NHAS bull., vol. 26, 1985) (artifacts: BLs)

Dartmouth College: (Hood Museum of Art, Hanover) (Proctor and Fox Collections) (Many areas of N.H.) (artifacts: Proctor: 250 items from all time periods, full range of tools) (SL-AV-PEA)

Davis, Paul: (PEA and personal) (Rye, Durham, and seacoast area) (artifacts: PO, ST, FL, CT, AS, CO, etc.)

Derry Historical Society: (DHS) (Derry) (artifacts: 1 PP)

Douglass, A. E.: (Smithsonian?) (N.H.) (artifacts: 6 GO, 1 KN, 2 AX, 102 PP, 1 GT, 1 PI)

Dowlin, Arthur: (?) (N.H.) (artifacts: some PP, stone club, utensils, BO)

Drake, A.: (different than site) (?) (N.H.) (artifacts: some PP, PO, and ST)

Durham Historical Association: (DHA) (N.H.) (artifacts: 20 PP, 1 WS, 3 GF) (SL-AV-PEA)

Edwards, Frank: (personal) (Manchester) (artifacts: 1 PE, 1 PL, 1 WS, 1 AT, 2 PP, 1 BL, FL)

Exeter Historical Society: (EHS) (Exeter) (artifacts: KN, PE)

Ferguson, Glenn: (William Fisher) (Pembroke, Manchester) (artifacts: BI, ET, ground ST—several of all of these) (Middle Archaic through Late Woodland)

Fitch, Earl: (personal) (Milford) (artifacts: 2 PE)

Fitch, Guy: (personal) (Milford) (artifacts: CT, 2 PE)

Folger, Homer: (?) (Newfields) (artifacts: Plano point)

Folsom, Charles: (BHS) (Boscawen area) (artifacts: 93 PP, 5 BL, 1 MO,
  1 WS, BS, BE, PI, 1 PL, 1 GO, 1 HO, 9 AX)

Fox, William (1827–1897): (DCM) (Lakes region) (artifacts: mostly GS
  tools: GO, MO, AX, BL)

Franklin Pierce College: (FPC) (Rindge) (artifacts: some)

Franklin Public Library: (FPL) (Franklin) (artifacts: 1 ST)

Goodale, John H.: (?) (N.H.) (artifacts: some)

Goodhue, Charles: (?) (Franklin area) (artifacts: many FL, 2 CO, PE,
  WS, 13 GO, 10 PO, 1 CT, 2 GF) (SL-AV-PEA)

Griffin Free Public Library and Museum: (GFPL&M) (Auburn)

Griffin, Sebastion (1831–1898): (?) (N.H.) (artifacts: some)

Hampton Historical Society: (Tuck Museum) (Hampton and Hampton
  Falls) (artifacts: PLs, MO & PE, CO, AD, PPs)

Hancock Historical Society: (HHS) (Hancock) (artifacts: MO,
  2 PE)

Harmon, Horace: (N.H. Hist. Soc.) (N.H. and OOS) (artifacts: 786
  pieces: PE, GO, PP, PO, SC, HS)

Hart: (St. Paul's School) (N.H.) (artifacts: 4 GO, UL, 1 EF, 1 WS,
  5 FL, 23 PP, 1 DR)

Harvard University: (Peabody Museum of Archaeology and Ethnology) (N.H. items) (artifacts: Neville site)

Hinsdale Historical Society: (HHS) (Hinsdale) (Smith Coll.)

Holmes, Paul: (PEA, PHS, personal) (Plaistow, Kingston, Newton
  Junction areas) (NHAS bull., vol. 23, 1982) (artifacts: numerous,
  from several sites, including PP, FL, ET, etc.)

Holt, H.: (?) (Milford area) (artifacts: 1 GO, 47 PP, 1 DR, 2 other)

Hooksett Historical Society: (HHS) (?)

House, James: (?) (Manchester area) (artifacts: some)

Huntoon: (?) (Henniker and OOS) (artifacts: N.H.: 1 GO, 1 PP)

Jewel, E. Perry: (GML) (Laconia and Lochmere) (artifacts: numerous,
  many PP, ST, PO)

Knight, William: (GML) (Laconia and Lakeport) (artifacts: several PP,
  etc.)

Kostegan, Eric: (personal) (Londonderry) (artifact: 1 PP)

Laconia Historical Society: (LHS) (Laconia)

Laconia Public Library, Gale Memorial: (GML) (Laconia) (Drake,
  Hoyt, Jewel, Knight, and Winchester Collections) (artifacts: see
  individual names)

Leadbeater and Barbour: Collections at Fryeburg Historical Society
  Museum, Fryeburg, Maine (some N.H. artifacts)

Libby, Henry: (Libby Museum, Wolfeboro) (Lakes region) (artifacts: several PP, PO, ST, and 2 CA)

Littleton Area Historical Society Museum: (LAHSM) (Littleton) (artifacts: small coll. of PP)

Lombardi, Frank: (?) (Plaistow) (artifacts: 1 GO)

Londonderry Historical Society Collections: (LHS) (Parmenter Farm Collection) (Londonderry) (artifacts: 4 PP, 3 ST, 1 QC); (H. Stevens Collection) (Manchester) (artifacts: 1 AX, GO, 3 PP, 1 PL, 1 AD, 2 BL, 1SA, 1 HS); (Webster/Beckley Collection) (Londonderry) (artifacts: 4 PP); (Londonderry Harvey Collection) (artifact: ST?)

Lund, Charles: (Nashua Historical Society) (Nashua) (NHAS bull., vol. 26, 1985) (artifacts: numerous PP, ST, FL, PO, etc.)

Manchester Historical Association Collections: (MHA) (Plaistow Historical Society) (PHS) (Newton Junction); (Berry, Clyde) (many areas) (See Berry); (Coleman; Kidder) (artifacts: PO, BE); (Kimball; Marshall, Harlan) (extensive); (Richardson) (artifacts: PP, X GO, PI, DR, BE, KN, MO, PE); (Smyth site) (numerous artifacts); Watson, Harry) (large); (Colby, Perley) (several artifacts); (Holmes, Paul) (several artifacts); (Shriker) (several artifacts); (Thebodeau, Ken) (numerous artifacts).

Matteuzzi, Eugene: (personal) (Derry) (artifacts: AD, BL)

Mayo, Bart: (Montshire Museum, Norwich, Vt.) (Lyme and OOS) (artifacts: PP, SC, numerous)

Michigan, University of: (U of M) (N.H.) (artifacts: 20 PP)

Mori, Richard: (personal) (Greenland, N.H.) (artifacts: 1 AX) (NHAS-SL-FI)

Mount Kearsarge Indian Museum: (MKIM) (Warner) (artifacts: numerous from both in and out of state)

Mount Washington Observatory Museum: (summit of Mt. Wash.) (Conn. River flood plain, N.H.) (donated to Museum by R. P. Goldthwait) (artifact: 1 ST)

Nashua Historical Society: (NHS) (Nashua) (Lund Coll.)

New Hampshire Antiquarian Society: (NHAS, Hopkinton, N.H.) (Collected by Dustin and Davis) (N.H.) (artifacts: some PP, PO, BO, BE, GS, CA)

New Hampshire Historical Society: (NHHS) (Concord) (artifacts: Harmon Collection)

Nisula: (personal) (Londonderry) (artifacts: 1 PP, SC, ST)

Odiorne Point State Park: (Seacoast Science Center) (Portsmouth) (artifacts: 60 PPs, PR)

Page, Charles: (?) (Concord area) (artifacts: 1 PO, PP)

Parker, Samuel: (Goodwin Lib.) (Farmington and OOS) (artifacts: 2 AX, 13 GO, 7 PE, 1 MO, 7 PO, 1 CT, 4 PL, 5 PP, 2 CT, 1 UL, 1 AT) (SL-AV-PEA)

Parsons, V. B.: (personal) (New Boston) (artifacts: 1 AX, 1 GO, 1 HO)

Peabody Essex Museum: (PEM, Salem, Mass.) (Has N.H. items)

Peabody Museum: (PM, Andover, Mass.) (Many N.H. items)

Plaistow Historical Society: (PHS) (Newton Junction) (Harvey Mitchell site artifacts) (NHAS bull., vol. 23, 1982); (BROX and Plaistow dump site) (several artifacts)

Plymouth State University: (PSU) (artifacts: see Sewall's Falls, Garvin's Falls, First Fort, New England Glassworks, Hazeltine Pottery, Squire Abiathar Britton)

Portsmouth Athenaeum: (PA) (N.H.) (G. Homey Collection 1824) (artifacts: 3 AX, GO, 2 PE, ST, steatite BO)

Portsmouth Public Library: (PPL) (Portsmouth) (artifacts: 104 PP)

Preston, Luther: (?) (Auburn) (artifacts: 25 items from Lake Massabesic area)

Reitan: (personal) (Hudson) (artifacts: 8 PP, 3 blanks, 3 CL PI, 3 CL marbles)

Saint Paul's School: (SPS) (N.H.) (Hart Collection)

Sandown Historical Society Museum: (SHSM) (Sandown) (artifacts: 1 BL, 1 SC)

Sargent, Howard: (DHR) (many areas of the state) (artifacts: numerous; all time periods)

Sargent Museum: (Howard Sargent Collection of books and artifacts; all time periods)

Seavey, Bruce: (personal) (Hooksett) (artifacts: 1 PE)

Shea, James: (personal) (Litchfield) (artifacts: 5 PP)

Silver, T.: (personal) (Boscawen) (artifacts: numerous)

Smith, John H.: (HHS) (Hinsdale) (artifacts: 320 PP, PO, KN, BL, PF, DR, 2 GO, PI, AT)

Squam Lake Science Center: (SLSC, Holderness) (artifacts: Goodhue Collection from the Franklin area)

Stevens, H.: (?) (Manchester) (NHAS-SL-FI) (artifacts: few)

Stotler, Christopher: (personal) (Warner and OOS) (artifacts: some prehist. and hist.)

Strawbery Banke Museum: (SBM) (Portsmouth) (artifacts: Hist.)

Sunapee Historical Society: (SHS) (Georges Mills) (artifacts: Russell's Inn site has 52 boxes)

Swanzey Historical Museum: (SHM) (Swanzey) (artifacts: 3 AX)

Thebodeau, Kenwood: (personal) (Litchfield) (artifacts: numerous items, PP, ST, etc.)

Thompson, Brownie: (E. McKenzie) (Conway area) (artifacts: numerous items, PP, PO, ST, etc.)

Thorndike: (personal) (Windham) (artifacts: 5 PP)

University of New Hampshire: (UNH, Durham) (artifacts: Rocks Road site, Smith Farm, Wadleigh Farm)

Viner, Brandon and Konnie: (personal) (Londonderry and OOS) (artifacts: 1 PP, 1 FL, and 1 BL [OOS])

Virgin, Charles: (personal) (Tilton) (artifacts: 1 unknown, found on his
own land)

Wells, Eugene Y.: (?) (NH) (artifacts: ST, etc.)

Wilcomb, Edgar and Charles: (?) (NH) (artifacts: some)

Wilmot Historical Society: (WHS) (Wilmot) (artifact: tomahawk)

Wilton, Town of: (Town hall) (Wilton) (artifact: 1 AX)

Winchester, Hoyt H.: (GM Lib., Laconia) (Lakes region) (artifacts:
60 PP, 5 PO, 6 DR, 4 GF, 2 blanks, 1 EF, 1 PI, 1 GO, 2 AD, BE)

Wingate, Charles E.: (son, J. Wingate) (Salem and OOS) (artifacts:
beautiful coll.; see OOS list)

Wolfeboro Historical Society: (WHS) (Wolfeboro)

Young, Laurence: (personal) (? and Londonderry) (artifacts: 1 AX and
6 polished round stones, maybe from ball mill, prehist. or hist.)

Out-of-State Collections Housed and/or Listed in Files

HOUSED AT PEA

Benton: (New York State)

Berry, C.: (Wyoming, Utah, and S. Dakota)

Clark, R.: (Connecticut)

Dewitt Co. Illinois Collection: (belongs to PEA) (Mississippian
artifacts)

North Carolina Collection: (pottery)

Ogunquit, Maine, Collection: (Turkey Brook)

Schulz: (Belongs to PEA) (Peoria, Ill.)

Tamburro: (Connecticut)

LISTED IN THE NHAS COLLECTIONS FILES

Berry, Clyde: (donated to NHAS in 1992 by J. Berry) (Wyoming and
Utah 1954 and rocks from S. Dakota) (artifacts: 43 PO, several PP,
ETs, SS)

Crosbie, Laurence: (some items OOS)

Douglass, A. E.: (Smithsonian?) (some items OOS)

Fruitland-Sears: (all collected in Mass.)

Ham, Marilyn: (personal, Londonderry, N.H.) (Collection from Lynn-
field, Mass., and the midwest)

Hannon, Horace: (N.H. Hist. Soc.) (partially from OOS)

Huntoon: (some from OOS)

Marble, A. D.: (owner: Bruce Dix, Exeter, N.H.) (collected in Hing-
ham, Salem, Newburyport, Mass.) (artifacts: 66 PP)

Mayo, Bart: (Montshire Museum, Norwich, Vt.) (some OOS)

Newcom, Lillian: (personal, Deerfield, N.H.) (collection from Kentucky: PP and PR)

Parker, Samuel: (Goodwin Lib., Farmington, N.H.) (some OOS)

Rhodes, Kenneth: (personal, Manchester, N.H.) (New York State)

Smith, Kenneth: (personal, Londonderry, N.H.) (artifacts: PP)

Smith, John H.: (Hinsdale Hist. Soc.) (some OOS)

Stotler, Christopher: (personal) (partially OOS)

Viner, Konnie: (personal, Londonderry, N.H.) (1 BL from Kansas)

Wingate, Charles E.: (Joseph Wingate, Londonderry, N.H.) (artifacts from Methuen, Lawrence, Haverhill, Andover, Boxford, Groveland, Mass., and Salem, N.H.) (filed under N.H.): 36 PP, 5 BL, 3 AX, 3 GO, 2 AD, 4 AT, 1 PI, 1 BE necklace, etc.) (Lovely collection!)

Part 2. Provenience of Artifacts by Towns

Allenstown: C. Berry Collection (MHA)

Amherst: C. Berry Collection (MHA)

Auburn: C. Berry Collection (MHA); D. Hardy Collection (PEA); P. Holmes Collection (PEA); Magnuson Collection

Bedford: C. Berry Collection (MHA)

Belmont: Drake site (Lochmere Archeological site) (DHR)

Berlin: Magnuson Collection, Mt. Jasper; R. Boisvert (DHR)

Boscawen: L. Adams Collection (BHS); First Fort site (PSU); C. Folsom Collection (BHS); Theo Silver Collection (P)

Bow: C. Berry Collection (MHA)

Brentwood: Smith Farm site

Canaan: C. Berry Collection (MHA); Historical Society Collection (CHS)

Canterbury: Canterbury Shaker Village (on site)

Claremont: Hunter site (H. Sargent)

Colebrook: V. Bunker and J. Potter (DHR)

Concord: C. Berry Collection (MHA); L. Corliss Collection (P); Garvin's Falls site (PEA) (PSU) (NHAS-SL-FI); Hart Collection (St. Paul's School); Hazeltine Pottery site (PSU); Sewall's Falls area (four sites) (PSU)

Conway: Chapman Collection

Deerfield: C. Berry Collection (MHA); V. Bunker (DHR)

Derry: W. Brown Collection (P); Historical Society Collection (DHSM); M. Proctor Collection (DCM); E. Matteuzzi Collection (P)

Dover: Museum Collection (DMC)

Durham: Adams Point site (PEA) (UNH); P. Davis Collection (P); Historical Association Collection (DHA)

Epsom: M. Yeaton site (PEA)

Exeter: A. Cote site; B. Dix Collection (P); Fort Rock site (PEA);
Historical Society Collection (EHS); Oaklands site; Pickpocket
Falls site (PEA) (NHAS-SL-FI); Powder House Point site (PEA);
Stanley site (PEA); W. White Collection (PEA on loan); J. Williams
Oaklands Site Collection (PEA) (NHAS-SL-FI)

Farmington: S. Parker Collection (Goodwin Library, Farmington)

Fitzwilliam: Meadow Pond site (PEA) (Sportman's Pond)

Franklin: C. Goodhue Collection; Proctor Collection (DCM); Public
Library Collection (FPL)

Freedom: V. Bunker (DHR); Hall site (DHR)

Georges Mills: Russell's Inn site (Sunapee Historical Society)

Goffstown: C. Berry Collection (MHA)

Gorham: Conner site (DHR)

Greenland: R. Mori Collection (P); NH40-5 site (NHAS-SL-FI)

Hampton: W. White Collections (PEA) (NHAS-SL-FI); Hunt's Island
site (DHR); Historical Society Collection (Tuck Museum)

Hampton Falls: Hampton Historical Society (Tuck Museum)

Hancock: Historical Society Collection (HHS)

Henniker: Huntoon Collection

Hinsdale: J. H. Smith Collection (HHS)

Holderness: Davison Brook site (DHR)

Hooksett: C. Berry Collection (MHA); B. Seavey (P)

Hudson: C. Berry Collection (MHA); Musquash Pond site (P. Hume);
Reitan Collection (P)

Jefferson: E. Boaras and P. Bock (DHR)

Kensington: Batchelder Collection; W. White Collection (PEA on
loan) (NHAS-SL-FI)

Kingston: Brox site (PHS); D. Hardy Collection (PEA); Tasha-Bodwell
Swamp site (DHR)

Laconia: Weirs Beach site; H. Winchester Collection (NHAS-SL-FI);
Gale Memorial Library Collection (GML); J. Demar Collection
(PEA)

Lee: Wadleigh Falls site (UNH); Wadleigh Farm site (PEA)

Litchfield: C. Berry Collection (MHA); Campbell site (DHR); Litchfield site; Rodonis Farm site (DHR); Smolt site (DHR); K. Thebodeau Collection (MHA); P. Colby Collection (MHA); Mill Pond
Estates (P. Hume); Chase Brook (LHS); J. Shea Collection (P)

Littleton: Historical Society Collection (LHSM)

Lochmere: Prindle Collection (PEA) (NHAS-SL-FI)

Londonderry: C. Berry Collection (MHA); E. Kostigan (P); Nisula
Collection (P); Parmenter Farm Collection (LHS); Webster/Beck-
ley Collection (LHS); Viner Collection (P); M. Harvey Collection
(LHS)

Lyme: Dennis Farm site (PEA and V. Pushee); B. Mayo Site Collection (P)

Lyndeboro: D. Hardy (Glass factory) (PEA)

Madison: West Branch Brook site (DHR)

Manchester: Belanger Collection (PEA); C. Berry Collections from
several areas (MHA) (PEA); Eddy site (PEA); F. Edwards Collection
(P); D. Hardy Collection (PEA); J. House Collection; H. Marshall
Collection (MHA); Neville site (NH38-5) (NHAS bull., vol. 18,
1975) (PMHU); G. Nicholas (Cohas Brook area) (PEA); K. Rhodes
Collection (PEA): Smyth site (NH38-4) (NHAS bull., vol. 18,
1975, and vol. 21, 1980) (PEA) (H. Sargent); H. Stevens Collection
(NHAS-SL-FI); G. Beaudoin Collection (P); B. Collins Collection
(PE); S. Johnson Collection (PEA)

Meredith: C. Berry Collection (MHA)

Merrimack: C. Berry Collection (MHA); Camp Naticook site (DHR);
Hume site (DHR); Sargent Road (DHR)

Milford: Contract near bridge and other sites (DHR); E. Fitch Collection (P); G. Fitch Collection (P); H. Holt Collection (?)

Milton: D. Chesley Collection (PEA)

Nashua: C. Berry Collection (MHA); several sites (DHR); C. Lund
Collection (NH52-1) (NHAS bull., vol. 26, 1985) (NHS); Mine
Falls site (PEA); F. Robinson Collection (P); Mine Falls Park site
(DHR)

New Boston: J. Dodge Collection (NH37-12) (NHAS bull., vol. 28,
1987) (P); V. B. Parsons Collection (P)

Newfields: L. Crosbie Collection (PEA); H. Folger Collection; White
and Finch (Stetson Farm site) (PEA) (NHAS-SL-FI)

New Hampton: C. Berry Collection (MHA)

Newmarket: Wiswall Falls site (PEA)

Newton Junction: Harvey Mitchell site (NH46-12) (NHAS bull., vol.
23, 1982); P. Holmes Collection (PEA and PHS)

Northfield: Lisbon Collection (PEA)

Northwood: site (NHAS-SL-FI)

Ossipee: R. Boisvert Field School sites (DHR); D. Hardy Collection
(PEA)

Pelham: C. Berry Collection (MHA)

Pembroke: C. Berry Collection (MEA); M. Blaisdell Collection (P);
G. Ferguson Collection (William Fisher); Mason site (DHR)

Pittsfield: Lisbon Collection (PEA)

Plaistow: Brox site (PHS); P. Holmes Collection (PEA); F. Lombardi
Collection; Plaistow Dump site (PHS)

Plymouth: R. Boyd Collection; NH19-2 site (NHAS-SL-FI)

Portsmouth: Athenaeum Collection (at PA); Strawbery Banke Collections (at SB); Public Library Collection (PPL); Collection from

Wentworth-Coolidge mansion grounds (Odiorne Pt. Seacoast Science Center)

Raymond: Raymond site (NHAS-SL-FI)

Rye: Wentworth Golf Course site (DHR) (PEA)

Salem: Mystery Hill site (on site)

Sandown: Sandown Historical Society Collection (SHSM)

Seabrook: Rocks Road site (NH47-21) (NHAS bull., vol. 28, 1987) (UNH) (NHAS-SL-FI)

Shelburne: Stevens Brook site (PEA)

South Hampton: South Hampton near Aspen Hill Collection (PEA) (NHAS-SL-FI)

Stratham: Wiggin site (DHR)

Swanzey: A. Whipple Collection (P); Whipple site; Historical Society Collection (SHS) and the Antiquarian Society of the Mount Caesar Union Library (location of library may be in Keene)

Tamworth: Argillite Workshop sites (NH20-1 and NH20-2) (NHAS bull., vol. 32, 1991) (UNH)

Temple: New England Glassworks (NHAS bull., vol. 27, 1986) (PSU and Temple Hist. Soc.)

Thornton's Ferry: Proctor Collection (DCM); Thornton's Ferry site (DHR)

Tilton: C. Berry Collection (MHA); E. Lisbon Collection (PEA); C. Virgin Collection (P); Lodge site (DHR); D. Howe (DHR)

Walpole: L. Chapin Collection (?)

Wilmot: Historical Society Collection (WHS)

Wilton: Town of Wilton Collection (WTH)

Winchester: Broad Brook site (NH41-17) (NHAS bull., vol. 25, 1984) (Colony House in Keene)

Windham: R. Barlow Collection (P); R. Thorndike Collection (P)

Wolfeboro: Historical Society collection (WHS); H. Libby Collection (Libby Museum in Wolfeboro); J. Wentworth Plantation site (NH27-2) (NHAS bull., vol. 30, 1989) (PSU)

Note: See PEA inventory sheet for site numbers

Part 3. Sites Described in *The New Hampshire Archeologist* since 1980, Listed by Town

Boscawen: "Preliminary Report of First Fort, Boscawen, NH (NH31-34)," Mary Dupre (vol. 26-1, 1985) (PSU) (hist.)

Brentwood: "A Report on the Smith Farm Site: Brentwood, NH," Chris DeCorse (vol. 22-1, 1981) (UNH) (prehist.); "The Homestead Once at North Road, Brentwood, New Hampshire," Martha E. Pinello (vol. 36-1, 1996) (hist.)

Canterbury: "Canterbury Shaker Village: Archeology and Landscape," David Starbuck (vol. 31-1, 1990) (on site) (hist.); "Recent Excavations at Canterbury Shaker Village," David Starbuck (vol. 37-1, 1997) (on site) (hist.); "Canterbury Shaker Village: Medicines as Seen Through Archeological Artifacts," Elizabeth B. Hall (vol. 43/44-1, 2003/2004) (on site) (hist.); "The Excavation of a Well at Canterbury Shaker Village," David Starbuck (vol. 43/44-1, 2003/2004) (on site) (hist.)

Charlestown: "The Water-Powered Mills of Charlestown, New Hampshire," Margery Reed and Virginia Moulton (vol. 37-1, 1997) (hist.)

Colebrook: "Early Occupation in the Far Upper Connecticut River Valley," Victoria Bunker and Jane Potter (vol. 39-1, 1999) (prehist.)

Concord: "Excavations at Sewall's Falls in Concord, NH (NH31- 30)," David Starbuck (vol. 23, 1982) (PSU) (prehist.); "Further Excavations at Sewall's Falls (NH31-30)," David Starbuck (vol. 25-1, 1984) (PSU) (prehist.); "Three Seasons of Site Survey and Excavations at Sewall's Falls (NH31-34)," David Starbuck (vol. 26-1, 1985) (PSU) (prehist.); "The Beaver Meadow Brook Site: Prehistory on the West Bank at Sewall's Falls, Concord, NH," Dennis Howe (vol. 19-1, 1988) (PSU) (prehist.); "The Garvin's Falls Site (NH37-1): The 1963–1970 Excavations by the NHAS," Eugene Winter (vol. 26-1, 1985) (PEA) (prehist.); "The Garvin's Falls Site (NH37-1): The 1982 Excavations," David Starbuck (vol. 26-1, 1985) (PSU) (prehist.); "Prehistoric Pottery of the Garvin's Falls Site," Victoria Kenyon (vol. 26-1, 1985) (PSU) (prehist.); "Argillite Blade Cache Found Near Sewall's Falls (NH31-30)," Alan Strauss (vol. 26-1, 1985) (owner: Less Corliss) (prehist.); "The World's Longest Timber Crib Dam: The Sewall's Falls Dam in Concord, New Hampshire," David Starbuck (vol. 26-1, 1985) (hist.); "The Hazeltine Pottery Site, Concord, NH (NH37-8)," David Starbuck and Mary Dupre (vol. 26-1, 1985) (PSU) (hist.); "Sewall's Falls Hydropower Station: Industrial Archeology Using Photographs," Dennis E. Howe (vol. 38-1, 1998) (hist.)

Cornish: "Sculpting Tools and Armature from an Atelier at Saint-Gaudens National Historic Site," James W. Mueller (vol. 37-1, 1997) (hist.)

Durham: "Jasper Flakes and Jack's Reef Points at Adams Point: Speculations on Interregional Exchange in Late Middle Woodland Times in Coastal New Hampshire," Howard M. Hecker (vol. 35-1, 1995) (prehist.)

Effingham: "The Thorne Site, 27-CA-26: A Late Paleoindian Site in East-Central New Hampshire," Richard A. Boisvert (vol. 43/44-1, 2003/2004) (DHR) (prehist.)

Epsom: "The Maurice Yeaton Farm Site (NH38-60)," Patricia Hume (vol. 32-1, 1991) (PEA) (prehist. and hist.).

Exeter: "The Stanley Site Revisited," Don Foster (vol. 23, 1982) (PEA) (prehist.).

Fitzwilliam: "Meadow Pond (NH49-1): Changing Site-Landform Associations at a Multiple-Component Site," George Nicholas (vol. 25-1, 1984) (prehist.)

Freedom: "SCRAP 1992 Excavations at the Hall Site, Freedom, NH," Steven L. Bayly (vol. 39-1, 1999) (prehist.); "Hornfels Tool-Making Industry in Freedom, NH, Site 27-CA-60," Victoria Bunker (vol. 42-1, 2002) (prehist.)

Gorham: "Early Contact Period Activity on a Great Thoroughfare: The Conner Site (27-CO-34)," Jane S. Potter (vol. 38-1, 1998) (hist. and prehist.)

Hampton Harbor: "The Hunt's Island Site: A Prehistoric Vantage Point on Hampton Harbor," Mark Greenly (vol. 39-1, 1999) (prehist.)

Holderness: "Defining the Dynamic Late Archaic Period at the Davison Brook Site, 27GR201," Robert Goodby (vol. 41-1, 2001) (prehist.)

Jefferson: "Recent Paleoindian Discovery: The First People in the White Mountain Region of New Hampshire," Edward F. Bouras and Paul M. Bock (DHR) (vol. 37-1, 1997) (prehist.)

Kingston: "The Tasha-Bodwell Swamp Site (27-RK-202), Rockingham County, New Hampshire," Douglas C. Kellogg and Robert G. Kingsley (vol. 39-1, 1999) (prehist.)

Lee: "Wadleigh Falls Island (NH39-1): A Preliminary Site Report," Laura Pope (vol. 22-1, 1981) (UNH) (prehist.); "The Wadleigh Falls Site (NH39-2): A Preliminary Report of the 1980 Excavations," David Skinas (vol. 22-1, 1981) (UNH) (prehist.)

Litchfield: "The Smolt Site: Seasonal Occupation in the Merrimack Valley," Victoria Kenyon (vol. 24, 1983) (DHR) (prehist.); "Two Woodland Components in Litchfield, NH," Victoria Bunker (vol. 29-1, 1988) (DHR) (prehist.)

Lyme: "The Dennis Farm Site: Late and Final Woodland Utilization of an Upland Location in the Connecticut River Drainage," Andrea Ohl (vol. 32-1, 1991) (artifacts are mostly housed at PEA) (prehist.).

Nashua: "The Lund Collection: Nashua, NH," Victoria Kenyon (vol. 26-1, 1985) (Nashua Hist. Soc.) (prehist.); "The Place Between: Archeology at the Mine Falls Park Site, Nashua, New Hampshire," Victoria Bunker and Jane Potter (vol. 36-1, 1996) (prehist.)

New Boston: "The Jim Dodge Site (NH37-12)," Jane Potter and Patricia Hume (vol. 28-1, 1987) (owner: Jim Dodge) (prehist.)

Newton Junction: "The Harvey Mitchell Site (NH46-2) Newton
Junction, NH," Paul Holmes (vol. 23, 1982) (Plaistow Hist. Soc. and
PEA) (prehist.)

Manchester: "Prehistoric Pottery at the Smyth Site," Victoria Kenyon
(vol. 22-1, 1981) (PEA) (prehist.); "Ancient Lifeways at the Smyth
Site (NH38-4)," Donald Foster, Victoria Kenyon, George Nicholas
II (vol. 22-2, 1981) (PEA) (prehist.)

New Castle: "Nautical Archeology in Hart's Cove," David C. Switzer
(vol. 33/34-1, 1994) (hist.)

Orford: "A Trash Pit behind the Squire Abiathar Britton House in
Orford," David Starbuck (vol. 40-1, 2000) (PSU) (hist.)

Plaistow: "Two Sites on Little River: The Plaistow Dump Site
(NH46-34) and the BROX/Galloway Site (NH46-35)," Patricia
W. Hume and Paul E. Holmes (vol. 38-1, 1998) (prehist.); "The
Paul H. Holmes Site, NH46-10: A Middle Archaic Site in Plaistow,
Rockingham County, New Hampshire," Paul E. Holmes and
Edward McKenzie (vol. 43/44-1, 2003/2004) (prehist.)

Portsmouth: "Using Historical Archeology to Rewrite the Myth of the
'Poor Widow': An Example from Nineteenth-Century Portsmouth,
New Hampshire," Kathleen Wheeler (vol. 35-1, 1995) (hist.); "The
History of Archeology at Strawbery Banke Museum," Mary Bentley
Dupre (vol. 35-1, 1995) (hist.); "The Historic Archeology of Deer
Street, Portsmouth, New Hampshire," Aileen B. Agnew (vol. 35-1,
1995) (hist.)

Seabrook: "A Preliminary Report on the Rocks Road Site (Seabrook
Station): Late Archaic to Contact Period Occupation in Seabrook,
NH," Brian Robinson and Charles Bolian (vol. 28-1, 1987) (UNH)
(prehist.); "Native American Ceramics from the Rocks Road Site,
Seabrook, New Hampshire," Robert G. Goodby (vol. 35-1, 1995)
(UNH) (prehist.)

Tamworth: "Argillite Workshops in Tamworth, NH," Robert Ewing
and Charles Bolian (vol. 32-1, 1991) (UNH) (prehist.)

Temple: "The New England Glassworks: New Hampshire's Boldest
Experiment in Early Glassmaking," David Starbuck (vol. 27-1,
1986) (PSU and Temple Hist. Soc.) (hist.)

Tilton: "Rescue Archeology at the Lodge Site, NH31-6-6," Justine B.
Gengras and Victoria Bunker (vol. 38-1, 1998) (prehist.); "A View
of Middle Archaic Life from Lithic Workshops," Dennis E. Howe
(vol. 40-1, 2000) (prehist.)

West Swanzey: "New Hampshire Paleo-Indian Research and the
Whipple Site," Mary Lou Curran (vol. 33/34-1, 1994)
(prehist.)

Winchester: "The Broad Brook Site (NH41-17), Pisgah State Park, NH," Faith Harrington (vol. 25-1, 1984) (Colony House Museum, Keene, NH) (hist.)

Wolfeboro: "America's First Summer Resort: John Wentworth's 18th-Century Plantation in Wolfeboro, New Hampshire," David Starbuck (vol. 30-1, 1989) (PSU) (hist.)

Part 4. New Hampshire Collections Housed at Private and Public Institutions

Colleges, Universities, and Schools
Dartmouth College: Hood Museum of Art, Hanover, N.H. (*see also* Museums)
Franklin Pierce College: Anthropology Dept., Rindge, N.H.
Harvard University: Peabody Museum, Cambridge, Mass. (N.H. items) (*see also* Museums)
Phillips Andover Academy: Peabody Museum, Andover, Mass. (N.H. items) (*see also* Museums)
Phillips Exeter Academy: Anthropology Dept. and Museum, Exeter, N.H. (*see also* Museums)
Plymouth State University: Dept. of Social Science, Plymouth, N.H.
St. Paul's School: Concord, N.H.
University of Michigan: Ann Arbor, Mich.
University of New Hampshire: Anthropology Dept., Durham, N.H.

Historical and Antiquarian Societies
Boscawen Historical Society
Canaan Historical Society
Concord: New Hampshire Historical Society (Museum of New Hampshire History)
Conway Historical Society
Derry Historical Society (Museum)
Durham Historical Association
Exeter Historical Society
Hampton Historical Society
Hancock Historical Society
Hinsdale Historical Society
Hooksett Historical Society
Hopkinton: New Hampshire Antiquarian Society (Museum)
Laconia Historical Society
Littleton Area Historical Society (Museum)
Londonderry Historical Society
Manchester Historical Association (Museum)

Nashua Historical Society (Museum)
Plaistow Historical Society (Museum)
Sandown Historical Society (Museum)
Sunapee Historical Society
Wilmot Historical Society
Wolfeboro Historical Society

Libraries
Auburn: Griffin Free Public Library and Museum
Berlin City Library
Farmington: Goodwin Library
Franklin Public Library
Kensington Library
Laconia: Gale Memorial Library
Portsmouth Athenaeum Library
Portsmouth Public Library
Swanzey: Antiquarian Society of the Mount Caesar Union Library

Museums
Andover, Mass.: Peabody Museum at Phillips Andover Academy (has
    N.H. items) (*see also* Colleges, Universities, and Schools)
Auburn: Griffin Museum (and Public Library)
Cambridge, Mass.: Peabody Museum at Harvard University (has N.H.
    items) (*see also* Colleges, Universities, and Schools)
Canterbury: Canterbury Shaker Village
Concord: New Hampshire Historical Society Museum
Conway: Conway Historical Society; Eastman-Lord House Museum
Derry: Derry Historical Society Museum
Dover: Woodman Institute Museum
Exeter: Phillips Exeter Academy Anthropology Museum (*see also* Col-
    leges, Universities, and Schools)
Hampton: Hampton Historical Society; Tuck Museum
Hanover: Hood Museum of Art, Dartmouth College (*see also* Colleges,
    Universities, and Schools)
Holderness: Squam Lakes Science Center
Hopkinton: New Hampshire Antiquarian Society Museum
Keene: Colony House Museum
Littleton: Littleton Area Historical Society Museum
Manchester: Manchester Historical Association Museum
Nashua: Nashua Historical Society Museum
North Conway: Mount Washington Observatory Museum, on top of
    Mount Washington
Plaistow: Plaistow Historical Society Museum

Portsmouth: Strawbery Banke Museum
Rye: Odiorne Point Science Center
Salem, Mass.: Peabody Essex Museum (has N.H. items)
Sandown: Sandown Historical Society Museum
Swanzey: Swanzey Historical Museum
Warner: Mount Kearsarge Indian Museum
Wolfeboro: Libby Museum

Archeological Sites with Museums
North Salem: America's Stonehenge (Mystery Hill)

Towns
Wilton: Town Hall

State
Concord: Division of Historical Resources, Archaeology Division
    (houses SCRAP projects and contract archeology collections)

Codes Used in this Compendium

Locations
DCM = Dartmouth College Museum (Hood Museum of Art)
DHR = Division of Historical Resources, Archaeology Division
GML = Gale Memorial Library, Laconia
HS = Historical Society
MHA = Manchester Historical Association
NHS = Nashua Historical Society
NHAS-SL-FI = New Hampshire Archeological Society Slide Files
OOS = Collected out of state
PMHU = Peabody Museum, Harvard University
P = Personal Collection
PEA = Phillips Exeter Academy
PHS = Plaistow Historical Society
PPL = Portsmouth Public Library
PSU = Plymouth State University
SHSM = Sandown Historical Society Museum
SB = Strawbery Banke
SL-AV-PEA = Slides Available at PEA
UNH = University of New Hampshire, Anthropology Department
WTH = Wilton Town Hall
WHS = Wolfeboro Historical Society
DMC = The Woodman Institute Museum, Dover

Artifacts

AD = adze
AS = abrading stone
AT = atlatl
AX = axe
BA = bannerstone
BE = bead
BL = blade
BO = bowl
CA = canoe
CO = core
CT = celt
DR = drill
EF = effigy
ET = edge tool
FL = flake
GO = gouge
GT = gorget
HS = hammerstone
HO = hoe

KN = knife
MA = maul
MC = miscellaneous
MO = mortar
PE = pestle
PI = pipe
PL = plummet
PO = pottery
PP = projectile point
PR = perforator
SA = shaft abrader
SC = scraper (edge tool also)
SR = stone rod
SS = spoke shave
ST = stone tool
TR = tightener
UL = ulu
WS = whetstone

Other Items of Interest

GF = gunflint
Hist. = historical
Prehist. = prehistoric
QC = quartz crystal

Appendix 3

New Hampshire Archeological Sites Listed on the National Register of Historic Places

The following is a listing of all properties in New Hampshire that have been entered onto the National Register of Historic Places principally or solely on the basis of their archeological significance. The year of listing is included in parentheses.

Belknap County

Endicott Rock, Weirs Channel, Laconia (1980)
Cultural affiliation: Early Archaic, Middle Archaic, Late Archaic

Lochmere Archeological District, Tilton (1982)
Cultural affiliation: Archaic, Woodland, American

The Weirs, Laconia (1975)
(Also known as NH26-1; Aquadoctan Archeological Site)
Cultural affiliation: Archaic, Woodland, English

Coos County

Mount Jasper Lithic Source, Berlin (1992)
(Also known as NH27-CO-1)
1½ miles northwest of confluence of Dead River and Androscoggin River
Cultural affiliation: Archaic, Woodland

Hillsborough County

New England Glassworks site, Temple (1975)
(Also known as Temple Glassworks Site)
Cultural affiliation: American

Merrimack County

Beaver Meadow Brook Archeological Site, Concord (1989)
(Also known as NH27-MR-3)
Cultural affiliation: Middle Archaic, Late Archaic, Middle Woodland

Sullivan County
   Hunter Archeological Site, Claremont (1976)
   (Also known as Hunter Farm; NH28-3)
   Cultural affiliation: Middle and Late Woodland, Iroquoian

While many archeological sites in New Hampshire have been determined to be eligible for the National Register, few have actually been listed. The standards of the National Register state that a property must demonstrate its "significance in American history, architecture, archeology, engineering, and culture ... and possess integrity of location, design, setting, materials, workmanship, feeling, and association." A property can also be significant under any or all of the following four criteria:

   A. ... associated with events that have made a significant contribution to the broad patterns of our history;
   or B. ... associated with the lives of persons significant in our past;
   or C. ... embody the distinctive characteristics of a type, period, or method of construction, or that represent the work of a master, or that possess high artistic values, or that represent a significant and distinguishable entity whose components may lack individual distinction;
   or D. ... have yielded or may be likely to yield, information important in prehistory or history.

It is oftentimes harder for an archeological site to convey its significance than it is for a well-documented historic structure. For example, to meet the requirements of criterion D, it is necessary to demonstrate that a property contains or is *likely* to contain information bearing on an important archeological research question. This requires considerable evidence, often obtained only through professional archeological excavations. When such a project is conducted, it is necessary to determine the time period, extent, stratification, and integrity of the site, as well as several other types of data. Also, an archeological site must not be completely excavated or it will no longer be eligible for the register. Nor is it always desirable for an archeological site to be listed, because many sites are located in areas where they cannot easily be protected. Thus to bring attention to the historical significance of an important prehistoric or historic archeological site could lead to its ultimate destruction.

Appendix 4

Selected Projectile Points from the Colby Collection

Illustrated by Ellen Pawelczak

This is a selection of projectile points from the Colby Collection, all of which are currently on display in the Anthropology Museum at Phillips Exeter Academy. Most or all of these were found in New Hampshire, although we do not know the exact provenience for most of them. As I indicated in the introduction, Solon Colby collected chiefly along the Merrimack River, especially at such prominent sites as the Smyth site, Sewall's Falls, and Garvin's Falls, so many of these artifacts were probably found in the vicinity of Manchester, Suncook, Bow, or Concord. These are neither all of the point types to be found in New Hampshire nor quite all of the point types in the Colby Collection. Still, they are a good representation of types from the Early Archaic (Eden-like, Kirk Stemmed, Bifurcate Base), Middle Archaic (Neville, Neville Variant, Stark), Late Archaic (Brewerton Eared Triangle, Atlantic, Normanskill, Small Stemmed, Susquehanna Broad), Early Woodland (Meadowood), Middle Woodland (Jack's Reef Corner-Notched), and Late Woodland (Levanna) periods. The Colby Collection does not include any examples of Clovis points.

All of the illustrations were prepared by Ellen Pawelczak, working from the original artifacts, and all are drawn at a scale of 1:1. In every case a photograph of the point is shown side by side with the illustration of that point.

FIG. APP.4.1. Eden-like point, 3 inches long, from Sewall's Falls.

FIG. APP.4.2. Kirk Stemmed point, 3¹⁄₁₆ inches long.

FIG. APP.4.3. Bifurcate Base point, 1¹⁵⁄₁₆ inches long.

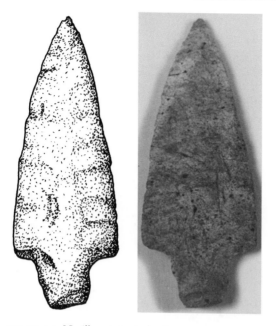

FIG. APP.4.5. Neville Variant point, 2¹¹/₁₆ inches long.

FIG. APP.4.4. Neville point, 3 inches long.

FIG. APP.4.7. Brewerton Eared Triangle point, 2³/₁₆ inches long.

FIG. APP.4.6. Stark point, 3⁹/₁₆ inches long.

FIG. APP.4.8. Atlantic point, 3⅝ inches long.

FIG. APP.4.9. Normanskill point, 2½ inches long.

FIG. APP.4.10. Small Stemmed point, 1⅜ inches long.

FIG. APP.4.11. Susquehanna Broad point, 2¹³/₁₆ inches long.

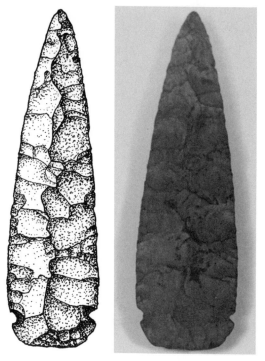

FIG. APP.4.12. Meadowood pont, 3⁷/₁₆ inches long, from Garvin's Falls.

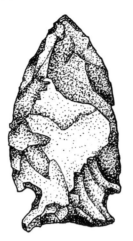

FIG. APP.4.13. Jack's Reef Corner-Notched point, 2³/₁₆ inches long.

FIG. APP.4.14. Levanna point, 1⅜ inches long.

Glossary

For additional terms pertaining to prehistoric sites, see the glossary in *The Archaeology of New England*, by Dean R. Snow (New York: Academic Press, 1980); for additional terms pertaining to historic archeological sites, see the glossary of terms in *The Great Warpath*, by David R. Starbuck (Hanover, N.H.: University Press of New England, 1999); and for additional terms pertaining to the broader field of archeology, see the glossary in *In the Beginning*, by Brian M. Fagan and Christopher R. DeCorse, 11th ed. (Upper Saddle River, N.J.: Pearson Prentice Hall, 2005).

assemblage: All of the artifacts found at a site, or within a component of a site.

atlatl: A spear thrower, used to lengthen the arm and to throw the spear straighter and further.

bannerstone: Another name for a winged atlatl weight, used to add heft to the spear thrower.

biface: A bifacially worked stone tool with flakes removed from both the upper and lower sides of the edge.

bulb of percussion: The bulge that appears on the surface of a flake of stone, typically with concentric rings that radiate out from the center of the point of impact.

channel flake: The flake that results from creating a flute on the side of a fluted point.

chert: A smooth, glassy stone used for making chipped-stone tools; the closest quarry sites exist in New York, Vermont, and Maine. Similar in appearance to flint but does not occur in chalk deposits.

Clovis Barrier: The moment in time when the first Clovis points began to be made; also the belief that no one lived in the New World before that time.

component: The occupation of an archeological site by a culture. A site may be single-component or multi-component.

contract archeology: See *cultural resource management*.

core: The lump of raw material from which a stone tool is manufactured.

creamware: A cream-colored earthenware that was developed by Thomas Astbury and Thomas Whieldon in the 1740s. Sets of dishes manufactured from this ware were popular in the second half of the eighteenth century and at the beginning of the nineteenth century.

cultural resource management (CRM): Studying and protecting the past for the benefit of all. This work is carried out in response to federal, state, and local acts and laws.

debitage: The debris left behind (flakes, chips, chunks) when manufacturing a stone tool.

delft: The same as tin-glazed earthenware.

flake: The stone that is removed from a core in the reduction process; essentially a waste product, but flakes were often utilized as sharp-edged tools. Every flake has a striking platform and a bulb of percussion.

flint: Chert that occurs naturally within chalk deposits; not found in New England.

fluted point: A Paleo-Indian projectile point that has had thinning flakes struck from one or both sides.

gunflint: A wedge- or prism-shaped blade of flint that produces a spark when struck against a steel frizzen, thereby firing a flintlock musket or pistol.

horticulture: Farming with simple hand tools, such as digging sticks or hoes; this is conducted on a smaller scale than agriculture, which produces a surplus to sell.

megafauna: The large Pleistocene animals that soon became extinct after the close of the glacial period.

midden: A garbage dump, showing evidence of human activity.

Moundbuilders: The Adena, Hopewell, and Mississippian peoples who initially built mounds for burial purposes and later built them as a base for temples and for worship.

palisade: A high fence or defensive wall of vertical posts set into the ground or into a ditch.

pearlware: An earthenware that evolved out of creamware in the 1770s but had a whiter glaze; a very common ware in the early nineteenth century but was phased out circa 1820.

pesthouse: A house used for quarantining patients, to keep those with contagious diseases away from the general public.

phase: The archeological evidence for a prehistoric culture, a grouping of cultural traits that can be defined in time and space.

pieces esquillees: Flakes or core fragments that were used to "split bone, wood, antler or ivory" (Snow 1980: 127).

plummet: A net sinker, used as a weight for net fishing.

polyethylene glycol (PEG): Wax used as a preservative in the conservation of wood, bone, and other organic materials. Prolonged soaking of an artifact in an aqueous solution of PEG causes the wax gradually to fill the pores and stabilize the material.

posthole/postmold: A posthole is the hole into which a post is placed or driven, whereas a postmold is what remains from the actual post that stood within the hole.

projectile point: A spear point, lance head, or arrowhead. The shape implies that it is meant to be thrown, but the edge needs to be examined to verify this.

radiocarbon dating: An absolute dating technique that uses the decay rate of carbon-14 to nitrogen to determine the age of a carbon-based object. The age is expressed in years before 1950 (B.P. = before present, or 1950).

reduction: The process of reducing a stone tool from a core down to the final desired shape (such as a projectile point).

redware: A red-bodied earthenware made from red clays, typically used for food preparation or storage throughout the colonial period and later.

roasting platform: An exceptionally large prehistoric firepit, denoting specialized ritual activity or large-scale food processing.

sachem: An Indian leader in New York or New England.

scratch blue: Incised decoration added to the surface of white salt-glazed stoneware. Cobalt, when added to these incisions, leaves a pattern of thin blue lines.

striking platform: The part of a flake that is struck so as to detach the flake from a larger mass of stone.

tin-glazed earthenware: An earthenware with a soft clay body and a glaze that contains tin oxide. This turns the ware white, such that it resembles porcelain; in use throughout the seventeenth and eighteenth centuries.

ulu: A semilunar knife of ground slate.

uniface: A unifacially worked stone tool, with flakes removed from only one side of the edge, as in the case of a scraper.

waster: An imperfect artifact spoiled in the process of manufacturing, as in the case of waster sherds or waster pipe fragments.

weir: A dam or fence of stakes or brush placed across a stream for the purpose of catching fish.

white salt-glazed stoneware: A refined English stoneware manufactured in molds throughout the eighteenth century.

whiteware: A hard white earthenware (such as "ironstone china") manufactured from about 1810 to the present.

wigwam: A rectangular bark-covered house with an arched roof.

Bibliography

Adovasio, James M., and Jake Page. 2002. *The First Americans: In Pursuit of Archaeology's Greatest Mystery*. New York: Random House.

Agnew, Aileen Button. 1985. "The Archeology of a Neighborhood: Deer Street, Portsmouth, New Hampshire." *Historical New Hampshire* 40(1–2): 72–83.

———. 1988. "Ceramics and the Sea Trade in Portsmouth, New Hampshire: 1765–1785." *Northeast Historical Archaeology* 17: 40–60.

———. 1989. *The Historic Archaeology of Deer Street, Portsmouth, New Hampshire*. Portsmouth: Institutional Archives, Strawbery Banke Museum.

———. 1995a. "The Historic Archeology of Deer Street, Portsmouth, New Hampshire: A Feature Presentation." *New Hampshire Archeologist* 35(1): 29–45.

———. 1995b. "Women and Property in Early 19th-Century Portsmouth, New Hampshire." *Historical Archaeology* 29(1): 62–74.

Anonymous. 1980–1981. "Pesthouse in Newmarket, New Hampshire." *Newsletter of the New Hampshire Archeological Society* (Dec. 1980–March 1981): 11–13.

Armstrong, John Borden. 1970. *Factory Under the Elms*. Cambridge, Mass.: M.I.T. Press; new ed., Phoenix Publishing, 1984.

Belcher, C. Francis. 1980. *Logging Railroads of the White Mountains*. Boston: Appalachian Mountain Club.

Belknap, Jeremy. 1792. "Monuments and Relics of the Indians." In *The History of New Hampshire*, by Jeremy Belknap, vol. 3. Boston: Belknap and Young.

Boisvert, Richard A. 1992. "The Mount Jasper Lithic Source, Berlin, New Hampshire: National Register of Historic Places Nomination and Commentary." *Archaeology of Eastern North America* 20: 151–166.

———. 1994a. "Piscataquog Fluted Point." *New Hampshire Archeological Society Newsletter* 10(2): 6–7.

———. 1994b. "Volunteerism in New Hampshire Archeology." *New Hampshire Archeologist* 33/34(1): 5–8.

———. 1998. "Israel River Complex: A Paleoindian Manifestation in Jefferson, New Hampshire." *Archaeology of Eastern North America* 26: 97–106.

———. 2003–2004. "The Thorne Site, 27-CA-26: A Late Paleoindian Site in East-Central New Hampshire." *New Hampshire Archeologist* 43/44(1): 19–43.

———. 2004. "Clovis Era Technology in Northern New Hampshire: The Israel River Complex." In *New Perspectives on First Americans Studies*, ed. Bradley T. Lepper and Robson Bonnichsen. College Station: Texas A&M Press. 49–54.

Boisvert, Richard A., and Gail N. Bennett. 2004. "Debitage Analysis of 27-HB-1, a Late Paleoindian/Archaic Stratified Site in Southern New Hampshire." *Archaeology of Eastern North America* 32: 89–100.

Bolian, Charles E. 1976–1977. "Weirs Beach: A Preliminary Report of the 1976 Excavations." *New Hampshire Archeologist* 19: 47–55.

———. 1980. "The Early and Middle Archaic of the Lakes Region, New Hampshire." In *Early and Middle Archaic Cultures in the Northeast: Occasional Publications in Northeastern Anthropology*, ed. David R. Starbuck and Charles E. Bolian, Rindge, N.H.: Department of Anthropology, Franklin Pierce College. 7: 115–134.

Bolian, Charles E., and Pamela J. Cressey. 1978. "A Tale of Two Parking Lots." In *Conservation Archaeology in the Northeast: Toward a Research Orientation*, ed. Arthur E. Spiess. Peabody Museum Bulletin 3. Cambridge, Mass.: Harvard University Press. 102–106.

Boswell, Mary Rose. 1994. "Documenting Laconia's Knitting Mills: A Comparison of the Belknap Mills Corporation and Two Present-Day Knitting Mills." *IA: The Journal of the Society for Industrial Archeology* 20(1–2): 32–49.

Bouras, Edward, and Paul Bock. 1997. "Recent Paleoindian Discovery: The First People in the White Mountain Region of New Hampshire." *New Hampshire Archeologist* 37(1): 70–76.

Bouton, Nathaniel. 1856. *The History of Concord, from its First Grant in 1725, to the Organization of the City Government in 1853, with a History of the Ancient Penacooks*. Concord, N.H.: B.W. Sanborn.

Bragdon, Kathleen J. 2001. *The Columbia Guide to American Indians of the Northeast*. New York: Columbia University Press.

Brown, Percy S. 1950. "The Work and Aims of the New Hampshire Archeological Society." Manuscript on file at the New Hampshire Historical Society, Concord, N.H. (Press release, February 6, 1950.)

Broyles, Bettye J. 1971. "Second Preliminary Report: The St. Alban's Site, Kanawha County, West Virginia." *West Virginia Geologic and Economic Survey, Report of Archaeological Investigations* 3. Morgantown, W.Va.

Bunker, Victoria. 1988. "Two Woodland Components in Litchfield, New Hampshire." *New Hampshire Archeologist* 29(1): 1–48.

———. 1992. "Stratified Components of the Gulf of Maine Archaic Tradition at the Eddy Site, Amoskeag Falls." *Occasional Publications in Maine Archaeology* 9: 135–148.

———. 1994. "New Hampshire's Prehistoric Settlement and Culture Chronology." *New Hampshire Archeologist* 33/34(1): 20–28.

———. 2002a. "Analysis and Interpretation of Early Ceramics from Sewalls and Amoskeag Falls, Merrimack River Valley, New Hampshire." In *A Lasting Impression: Coastal, Lithic, and Ceramic Research in New England Archaeology*, ed. Jordan E. Kerber. Westport, Conn.: Praeger. 207–222.

———. 2002b. "Hornfels Tool-Making Industry in Freedom, NH, Site 27-CA-60." *New Hampshire Archeologist* 42(1): 1–56.

Bunker, Victoria, and Jane Potter. 1999. "Early Occupation in the Far Upper Connecticut River Valley." *New Hampshire Archeologist* 39(1): 70–81.

Burtt, J. Frederic. 1971. "Solon Baker Colby, 1900–1971. President Emeritus, N.H.A.S." *New Hampshire Archeologist* 16: unpaginated.

Caduto, Michael J. 2003. *A Time Before New Hampshire: The Story of a Land and Native Peoples*. Hanover, N.H.: University Press of New England.

Caldwell, Joseph R. 1958. *Trend and Tradition in the Prehistory of the Eastern United States*. Menasha, Wis.: American Anthropological Association Memoir 88.

Calloway, Colin G. 1991. *Dawnland Encounters: Indians and Europeans in Northern New England*. Hanover, N.H.: University Press of New England.

Candee, Richard M. 1970. "Merchant and Millwright: The Water-Powered Sawmills of the Piscataqua." *Old-Time New England* 60(4): 131–149.

———. 1985. *Atlantic Heights: A World War I Shipbuilders' Community*. Portsmouth, N.H.: Peter E. Randall for the Portsmouth Marine Society.

Cassedy, Daniel Freeman. 1984. "The Spatial Structure of Lithic Reduction: Analysis of a Multicomponent Prehistoric Site in the Lakes Region of New Hampshire." M.A. thesis, Department of Anthropology, SUNY Binghamton.

———. 1991. *A Prehistoric Inventory of the Upper Connecticut River Valley*. Raleigh, N.C.: Privately printed.

Cassedy, Daniel Freeman, and Kimberly Parson. 2003. "Phase 1-B Archaeological Survey Upgrade of a Portion of Route 2 in Jefferson & Randolph, Coos County, New Hampshire. NHS-X-034-1(18), 13602." Prepared by URS Corporation for the New Hampshire Department of Transportation, Concord, N.H.

Chapman, Jefferson. 1975. *The Rose Island Site*. Knoxville: University of Tennessee, Department of Anthropology, Report of Investigations 14.

———. 1977. *Archaic Period Research in the Lower Little Tennessee River Valley*. Knoxville: University of Tennessee, Department of Anthropology, Report of Investigations 18.

Chatters, James C. 2001. *Ancient Encounters: Kennewick Man and the First Americans*. New York: Simon & Schuster.

Chesley, W. Dennis. 1978–1979. "Pioneers in New Hampshire Archaeology: Sebastian S. Griffin." *New Hampshire Archeologist* 20: 70–72.

———. 1982. "Pioneers in New Hampshire Archaeology: Samuel Sewall Parker." *New Hampshire Archeologist* 23: 128–130.

Chesley, W. Dennis, and Mary Beth McAllister. 1981. "Pioneers in New Hampshire Archaeology: Wentworth Cheswill, Esquire." *New Hampshire Archeologist* 22(1): 73–79.

Coe, Joffre L. 1964. "Formative Cultures of the Carolina Piedmont." *Transactions of the American Philosophical Society* 54(5).

Colby, Solon B. 1975. *Colby's Indian History: Antiquities of the New Hampshire Indians and Their Neighbors*. Center Conway, N.H.: Walker's Pond Press.

Cook, Sherburne F. 1976. *The Indian Population of New England in the Seventeenth Century*. Berkeley: University of California Press.

Cotter, John L. 1958. *Archeological Excavations at Jamestown Colonial National Historical Park and Jamestown National Historic Site, Virginia*. Archeological Research Series 4. Washington, D.C.: National Park Service.

———. 1977. "Continuity in Teaching Historical Archaeology." In *Teaching and Training in American Archaeology: A Survey of Programs and Philosophies*, ed. William P. McHugh. University Museum Studies, no. 10. Carbondale: University Museum, Southern Illinois University at Carbondale. 100–107.

Curran, Mary Lou. 1980. *Studying Human Adaptation at a Paleo-Indian Site: A Preliminary Report*. Research Report 18. Amherst: University of Massachusetts.

———. 1984. "The Whipple Site and Paleoindian Tool Assemblage Variation: A Comparison of Intrasite Structuring." *Archaeology of Eastern North America* 12: 5–24.

———. 1987. "The Spatial Organization of Paleoindian Populations in the Late Pleistocene of the Northeast." Ph.D. dissertation, University of Massachusetts, Amherst.

———. 1994. "New Hampshire Paleo-Indian Research and the Whipple Site." *New Hampshire Archeologist* 33/34(1): 29–52.

Day, Gordon M. 1978. "Western Abenaki." In vol. 15, *Northeast*, ed. Bruce G. Trigger. *Handbook of North American Indians*, 148–159. Washington, D.C.: Smithsonian Institution Press.

DeCorse, Chris R. 1978–1979. "Analysis of Feature 6 at the Marshall Pottery Site." *New Hampshire Archeologist* 20: 31–48.

Deetz, James. 1996. *In Small Things Forgotten*. New York: Anchor Books.

DeLony, Eric. 1993. "Surviving Cast- and Wrought-Iron Bridges in America." *IA: The Journal of the Society for Industrial Archeology* 19(2): 17–47.

Derry, Linda, and Marley R. Brown III. 1987. "Excavation at Colonial Williamsburg Thirty Years Ago: An Archeological Analysis of Cross-Trenching behind the Peyton Randolph Site." *American Archeology* 6(1): 10–19.

Dillehay, Thomas D. 1989. *Monte Verde: A Late Pleistocene Settlement in Chile*, vol. 1, *Paleoenvironment and Site Context*. Washington, D.C.: Smithsonian Institution Press.

———. 1997. *Monte Verde: A Late Pleistocene Settlement in Chile*, vol. 2, *The Archaeological Context and Interpretation*. Washington, D.C.: Smithsonian Institution Press.

Dincauze, Dena F. 1968. "Cremation Cemeteries in Eastern Massachusetts." *Papers of the Peabody Museum of Archaeology and Ethnology, Harvard University* 59(1).

———. 1971. "An Archaic Sequence for Southern New England." *American Antiquity* 36(2): 194–198.

———. 1972. "The Atlantic Phase: A Late Archaic Culture in Massachusetts." *Man in the Northeast* 4: 40–61.

———. 1975. "The Neville Site: 8,000 Years at Amoskeag." *New Hampshire Archeologist* 18: 2–4.

———. 1976. *The Neville Site: 8,000 Years at Amoskeag*. Peabody Museum Monographs 4. Cambridge, Mass.: Peabody Museum of Archaeology and Ethnology, Harvard University.

Dincauze, Dena F., and Mitchell T. Mulholland. 1977. "Early and Middle Archaic Site Distributions and Habitats in Southern New England." In *Amerinds and Their Paleoenvironments in Northeastern North America*, ed. Walter Newman and Bert Salwen. Annals of the New York Academy of Sciences 288. New York: The New York Academy of Sciences. 439–456.

Dublin, Thomas. 1979. *Women at Work*. New York: Columbia University Press.

Dupre, Mary B. 1985a. "Preliminary Report of First Fort, Boscawen, N.H. (NH31-34)." *New Hampshire Archeologist* 26(1): 117–125.

———. 1985b. "Searching for New Hampshire Redware Potters." *Historical New Hampshire* 40(1–2): 47–60.

———. 1995. "The History of Archeology at Strawbery Banke Museum." *New Hampshire Archeologist* 35(1): 12–28.

Eastman, David. 1974. "Archeological Dig." *New Hampshire Magazine* (October): 2–4.

Edwards, Diana, Steven R. Pendery, and Aileen Button Agnew. 1988. "Generations of Trash: Ceramics from the Hart-Shortridge House, 1760–1860, Portsmouth, New Hampshire." *American Ceramic Circle Journal* 6: 29–51.

Ewing, Robert, and Charles Bolian. 1991. "Argillite Workshops in Tamworth, New Hampshire." *New Hampshire Archeologist* 32(1): 87–95.

Fagan, Brian M. 2005. *Ancient North America.* 4th ed. New York: Thames & Hudson.

Feder, Kenneth L. 2006. *Frauds, Myths, and Mysteries.* 5th ed. New York: McGraw-Hill.

Feldman, Mark. 1976. *Mystery Hill.* North Salem, N.H.: Mystery Hill Press.

Fell, Barry. 1989. *America B.C.* Rev. ed. New York: Pocket Books.

Finch, Eugene D. 1967. "The Stanley Site, NH47-18." *New Hampshire Archeologist* 14: 13–16.

———. 1969. "The Great Bay Site." *New Hampshire Archeologist* 15: 1–13.

———. 1971. "The Litchfield Site: A Preliminary Report." *New Hampshire Archeologist* 16: 1–15.

Fonda, Christine. 1994. "New Hampshire IA Sites on the National Register of Historic Places." *IA: The Journal of the Society for Industrial Archeology* 20(1–2): 152–164.

Ford, Barbara, and David C. Switzer. 1982. *Underwater Dig: The Excavation of a Revolutionary War Privateer.* New York: William Morrow and Co.

Foster, Donald W. 1982. "The Stanley Site Revisited." *New Hampshire Archeologist* 23: 37–63.

Foster, Donald W., Victoria B. Kenyon, and George P. Nicholas II. 1981. "Ancient Lifeways at the Smyth Site, NH 38-4." *New Hampshire Archeologist* 22(2): 1–91.

Freeman, Rodney, and Katherine C. Donahue. 1994. "The Draper-Maynard Sporting Goods Company of Plymouth, New Hampshire, 1840–1937." *IA: The Journal of the Society for Industrial Archeology* 20(1–2): 139–151.

Funk, Robert E. 2004. "An Ice Age Quarry-Workshop: The West Athens Hill Site Revisited." *New York State Museum Bulletin* 504.

Funk, Robert E., and Charles F. Hayes III, eds. 1977. *Current Perspectives in Northeastern Archeology: Essays in Honor of William A. Ritchie.* Researches and Transactions of the New York State Archeological Association 17(1). Rochester, N.Y.: New York State Archeological Association.

Funk, Robert E., and David W. Steadman. 1994. *Archaeological and Paleoenvironmental Investigations in the Dutchess Quarry Caves, Orange County, New York.* Persimmon Press Monographs in Archaeology. Buffalo: Persimmon Press.

Garvin, Donna-Belle. 1994. "The Granite Quarries of Rattlesnake Hill: The Concord, New Hampshire, 'Gold Mine.'" *IA: The Journal of the Society for Industrial Archeology* 20(1–2): 50–68.

Garvin, James L. 1994. "Small-Scale Brickmaking in New Hampshire." *IA: The Journal of the Society for Industrial Archeology* 20(1–2): 19–31.

Garvin, James L., and Donna-Belle Garvin. 1985. *Instruments of Change: New Hampshire Hand Tools and Their Makers, 1800–1900*. Canaan, N.H.: Phoenix Publishing for the New Hampshire Historical Society.

Gengras, Justine B. 1996. "Radiocarbon Dates for Archeological Sites in New Hampshire." *New Hampshire Archeologist* 36(1): 8–15.

Gengras, Justine B., and Victoria Bunker. 1998. "Rescue Archeology at the Lodge Site, NH31-6-6." *New Hampshire Archeologist* 38(1): 1–33.

Goodby, Robert G. 1995. "Native American Ceramics from the Rock's Road Site, Seabrook, New Hampshire." *New Hampshire Archeologist* 35(1): 46–60.

———. 2001. "Defining the Dynamic Late Archaic Period at the Davison Brook Site, 27GR201." *New Hampshire Archeologist* 41(1): 1–87.

Goodwin, William B. 1946. *The Ruins of Greater Ireland in New England*. Boston: Meador Press.

Gramly, Richard Michael. 1980. "Prehistoric Industry at the Mt. Jasper Mine, Northern New Hampshire." *Man in the Northeast* 20: 1–23.

———. 1982. "The Vail Site: A Palaeo-Indian Encampment in Maine." *Bulletin of the Buffalo Society of Natural Sciences* 30.

———. 1984. "Kill Sites, Killing Ground and Fluted Points at the Vail Site." *Archaeology of Eastern North America* 12: 110–121.

———. 1988. *The Adkins Site: A Palaeo-Indian Habitation and Associated Stone Structure*. Buffalo: Persimmon Press.

Gramly, Richard Michael, and Stephen L. Cox. 1976. "A Prehistoric Quarry-Workshop at Mt. Jasper, Berlin, N.H." *Man in the Northeast* 11: 71–74.

Grimes, John R., W. Eldridge, B. G. Grimes, A. Vaccaro, F. Vaccaro, J. Vaccaro, N. Vaccaro, and A. Orsini. 1984. "Bull Brook II." *Archaeology of Eastern North America* 12: 159–183.

Gross, Laurence F. 1988. "Building on Success: Lowell Mill Construction and Its Results." *IA: The Journal of the Society for Industrial Archeology* 14(2): 23–34.

Hadingham, Evan. 2004. "America's First Immigrants." *Smithsonian* 35(8): 90–98.

Hamilton, D. L., and Robyn Woodward. 1984. "A Sunken 17th-Century City: Port Royal, Jamaica." *Archaeology* 37(1): 38–45.

Hareven, Tamara K., and Randolph Langenbach. 1978. *Amoskeag: Life and Work in an American Factory City*. Hanover, N.H.: University Press of New England.

Harrington, Faith. 1983a. "Follett Site Reveals Puddle Dock Maritime Role." *Strawbery Banke Newsletter* (April 1983): 1, 6.

———. 1983b. "Strawbery Banke: A Historic Waterfront Neighborhood." *Archaeology* 36(3): 52–59.

———. 1984. "The Broad Brook Site (NH41-17), Pisgah State Park, New Hampshire." *New Hampshire Archeologist* 25(1): 12–30.

———. 1985a. "Archeological Testing at the Original Location of Fort No. 4, Charlestown, N.H. (NH34-4)." *New Hampshire Archeologist* 26(1): 127–134.

———. 1985b. "Sea Tenure in Seventeenth-Century New England: Native Americans and Englishmen in the Sphere of Marine Resources, 1600–1630." Ph.D. dissertation, University of California, Berkeley.

———. 1985c. "Sea Tenure in Seventeenth-Century New Hampshire: Native Americans and Englishmen in the Sphere of Coastal Resources." *Historical New Hampshire* 40(1–2): 18–33.

———. 1989. "The Emergent Elite in Early 18th Century Portsmouth Society: The Archaeology of the Joseph Sherburne Houselot." *Historical Archaeology* 23(1): 2–18.

———. 1992. "Deepwater Fishing from the Isles of Shoals." In *The Art and Mystery of Historical Archaeology: Essays in Honor of James Deetz*, ed. Anne E. Yentsch and Mary C. Beaudry. Boca Raton: CRC Press. 249–266.

Harrington, Faith, and Victoria B. Kenyon. 1987. "New Hampshire Coastal Sites Survey, Summer 1986." *New Hampshire Archeologist* 28(1): 52–62.

Harrington, Jean C. 1952. "Historic Site Archeology in the United States." In *Archeology of Eastern United States*, ed. James B. Griffin. Chicago: University of Chicago Press. 335–344.

Hart, John P., and Christina B. Rieth, eds. 2002. *Northeast Subsistence-Settlement Change, A.D. 700–1300*. New York State Museum Bulletin 496.

Hecker, Howard M. 1981. "Preliminary Physical Anthropological Report on the 650-Year-Old Skeleton from Seabrook, New Hampshire." *Man in the Northeast* 21: 37–60.

Hencken, Hugh. 1939. The 'Irish Monastery' at North Salem, N.H." *New England Quarterly* 12: 429–442.

Hewes, Robert. 1781. Advertisements in the *Boston Gazette* and the *Country Journal*. Oct. 29, Nov. 5, and Nov. 9.

Holmes, Paul. 1972. "The New Hampshire Archeological Society: The First Quarter-Century." *New Hampshire Archeologist* 17: 4–6.

Howe, Antonett. 1989. "Structure 3." *New Hampshire Archeologist* 30(1): 79–97.

Howe, Dennis E. 1988. "The Beaver Meadow Brook Site: Prehistory on the West Bank at Sewall's Falls, Concord, New Hampshire." *New Hampshire Archeologist* 29(1): 49–107.

———. 1993. "The Page Belting Company: A Study of 19th-Century Power Transmission Belting Manufacturing." *IA: The Journal of the Society for Industrial Archeology* 19(1): 5–20.

———. 1994a. "Industrial Archeology: A Survey of Research in New Hampshire." *New Hampshire Archeologist* 33/34(1): 105–113.

———. 1994b. "The Maintenance of New Hampshire's First Polyphase Hydroelectric Station." *IA: The Journal of the Society for Industrial Archeology* 20(1–2): 119–138.

———. 2000. "A View of Middle Archaic Life from Lithic Workshops." *New Hampshire Archeologist* 40(1): 1–42.

Hume, Gary W. 1991. "Joseph Laurent's Intervale Camp: Post-Colonial Abenaki Acculturation and Revitalization in New Hampshire." In *Algonkians of New England: Past and Present*, ed. Peter Benes. Dublin Seminar for New England Folklife Annual Proceedings 1991. Boston: Boston University Press. 101–113.

———. 2003–2004. "Piecing Together My Past: Reflections on Being and Becoming State Archaeologist." *New Hampshire Archeologist* 43/44(1): 1–11.

Hume, Patricia, and Donald Foster. 1994. "A Guide to New Hampshire Sites and Collections." *New Hampshire Archeologist* 33/34(1): 114–126.

Ingersoll, Dan W. 1971a. "Problems of Urban Historical Archeology." *Man in the Northeast* 2: 66–74.

———. 1971b. "Settlement Archaeology at Puddle Dock." Ph.D. dissertation, Harvard University.

Jackson, Donald C. 1988. *Great American Bridges and Dams.* Washington, D.C.: Preservation Press.

Jordan, Douglas F. 1960. "The Bull Brook Site in Relation to 'Fluted Point' Manifestation in Eastern North America." Ph.D. dissertation, Harvard University.

Kelly, A. R. 1939. "The Macon Trading Post, an Historical Foundling." *American Antiquity* 4(4): 328–333.

Kelso, Gerald K., and Faith Harrington. 1989. "Pollen Record Formation Processes at the Isles of Shoals: Botanical Records of Human Behavior." *Northeast Historical Archaeology* 18: 70–84.

Kenyon, Victoria B. 1981. "Prehistoric Pottery at the Smyth Site." *New Hampshire Archeologist* 22(1): 31–48.

———. 1982. "Middle Woodland Pottery of the Central Merrimack Valley in N.H." *New Hampshire Archeologist* 23: 103–118.

———. 1983. "River Valleys and Human Interaction: A Critical Evaluation of Middle Woodland Ceramics in the Merrimack River Valley." Ph.D. dissertation, Boston University.

———. 1985a. "Prehistoric Pottery of the Garvin's Falls Site." *New Hampshire Archeologist* 26(1): 43–60.

———. 1985b. "The Prehistoric Pottery of the Smyth Site." In *Ceramic Analysis in the Northeast*, ed. James B. Petersen. Occasional Publications in Northeastern Anthropology 9. Rindge, N.H. 89–107.

———. 1986. "Eddy Site Radiocarbon Dates." *New Hampshire Archeological Society Newsletter* 2(2): 4.

Kenyon, Victoria B., and Donald W. Foster. 1980. "The Smyth Site (NH38-4): Research in Progress." *New Hampshire Archeologist* 21: 44–54.

Keyworth, William Gordon. 1973. *The Covered Bridges of New Hampshire.* Wentworth, N.H.: Scenes of New England.

Laurent, Stephen, and Father Joseph Aubery. 1995. *French Abenaki Dictionary.* Portland, Maine: Chisholm Bros. Publishers/Grand Trunk Publishers.

Levine, Mary Ann, Kenneth E. Sassaman, and Michael S. Nassaney, eds. 1999. *The Archaeological Northeast.* Westport, Conn.: Bergin & Garvey.

Lewandoski, Jan Leo. 1990. "The Restoration of the Cornish-Windsor Bridge." *Society for Industrial Archeology—New England Chapters Newsletter* 10(1): 11–16.

Lottero, Patricia A. 1983. *New Hampshire Indians: A Guide for Instruction.* Occasional Publication, Institute for New Hampshire Studies, Plymouth State College. Plymouth, N.H.: Plymouth State College.

Lyford, James O., ed. 1903. *History of Concord, New Hampshire.* Concord, N.H.: Rumford Press.

Malone, Joseph J. 1964. *Pine Trees and Politics: The Naval Stores and Forest Policy in Colonial New England, 1691–1775*. Seattle: University of Washington Press.

Marshall, Harlan A. 1942. "Some Ancient Indian Village Sites Adjacent to Manchester, New Hampshire." *American Antiquity* 7: 359–363.

Mayer, John. 1994. "The Mills and Machinery of the Amoskeag Manufacturing Company of Manchester, New Hampshire." *IA: The Journal of the Society for Industrial Archeology* 20(1–2): 69–79.

Maymon, Jeffrey H., and Charles E. Bolian. 1992. "The Wadleigh Falls Site: An Early and Middle Archaic Period Site in Southeastern New Hampshire." *Occasional Publications in Maine Archaeology* 9. Augusta, Maine: Maine Historic Preservation Commission. 117–134.

McKenzie, Ed. 2001. "On the Death of Stephen Laurent, Longtime NHAS Member and Truly an 'Indian Archeologist.'" *New Hampshire Archaeological Society Newsletter* 17(2): 5–6.

Meyer, Karl Ernest. 1973. *The Plundered Past*. Atheneum.

Moeller, Roger W. 1980. *6LF21: A Paleo-Indian Site in Western Connecticut*. Occasional Paper Number 2. Washington, Conn.: American Indian Archaeological Institute.

Moorehead, Warren King. 1930. "Explorations in the Merrimack Valley." *Phillips Academy Bulletin* 24: 12–14.

———. 1931. *The Merrimack Archaeological Survey: A Preliminary Paper, with Supplementary Notes by Benjamin L. Smith on the Concord Valley*. Salem, Mass.: Peabody Museum.

New England: An Inventory of Historic Engineering and Industrial Sites. 1974. Washington, D.C.: The Historic American Engineering Record.

Openo, Woodard D. 1988. *The Sarah Mildred Long Bridge: A History of the Maine–New Hampshire Interstate Bridge from Portsmouth, New Hampshire, to Kittery, Maine*. Portsmouth: Peter E. Randall.

Orser, Charles E., Jr., and Brian M. Fagan. 1995. *Historical Archaeology*. New York: HarperCollins College Publishers.

Pave, Marvin. 1985. "Group Going to Falklands to Save Last Clipper Ship." *Boston Globe* (September 29): 65.

Pendery, Steven R. 1978. "Urban Process in Portsmouth, New Hampshire: An Archeological Perspective." In *New England Historical Archeology*, ed. Peter Benes. Dublin Seminar for New England Folklife: Annual Proceedings, 1977. Boston: Boston University Scholarly Publications. 24–35.

———. 1981. "Summary Report: Marshall/Toogood Sites Development Project." Manuscript on file at the Jones House Archaeology Center, Strawbery Banke Museum, Portsmouth, N.H.

———. 1984. "The Archeology of Urban Foodways in Portsmouth, New Hampshire." In *Foodways in the Northeast*, ed. Peter Benes. Dublin Seminar for New England Folklife: Annual Proceedings, 1982. Boston: Boston University Scholarly Publications. 9–27.

———. 1985. "Changing Redware Production in Southern New Hampshire." In *Domestic Pottery of the Northeastern United States, 1625–1850*, ed. Sarah Peabody Turnbaugh. Orlando: Academic Press. 101–118.

Penn, Theodore Z., and Roger Parks. 1975. "Nichols-Colby Sawmill in Bow, New Hampshire." *IA: The Journal of the Society for Industrial Archeology* 1(1): 1–12.

Petersen, James B., and David E. Putnam. 1992. "Early Holocene Occupation in the Central Gulf of Maine Region." *Occasional Publications in Maine Archaeology* 9. Augusta, Maine: Maine Historic Preservation Commission. 13–61.

Pilkovsky, Ann M., and Richard Boisvert. 2004. "New Hampshire Division of Historical Resources Archaeology Bureau, Site Form for DHR Site No. 27-CA-15." On file in the New Hampshire Division of Historical Resources, Concord, N.H.

Pinello, Martha E. 1989. "Archaeological Formation Processes and Household Boundaries at Four Domestic Lots in the North End of Portsmouth, New Hampshire, 1730–1830." M.A. thesis, University of Massachusetts, Boston.

Poole, Claire. 2001. "Challenging the Clovis Paradigm." *American Archaeology* 5(3): 22–27.

Pope, Laura. 1981. "Wadleigh Falls Island NH 39-1: A Preliminary Site Report." *New Hampshire Archeologist* 22(1): 8–15.

Potter, Chandler E. 1856. *The History of Manchester.* Manchester, N.H.: C. E. Potter.

Potter, Jane S. 1998. "Early Contact Period Activity on a Great Thoroughfare: The Conner Site (27-CO-34)." *New Hampshire Archeologist* 38(1): 52–66.

Potter, Parker, and David Switzer. 1989. *Historic Context 65: SHIPWRECKS.* Concord, N.H.: N.H. Division of Historical Resources.

Putnam, Frederick W. 1873. "Description of a Carved Stone Representing a Cetacean, Found at Seabrook, N.H." *Essex Institute Bulletin* 5: 111–114.

Rieth, Christina B. 2002. Introduction. In *Northeast Subsistence-Settlement Change A.D. 700–1300,* ed. John P. Hart and Christina B. Rieth. New York State Museum Bulletin 496. Albany, N.Y. 1–10.

Ritchie, William A. 1932. *The Lamoka Lake Site: The Type Station of the Archaic Algonkin Period in New York.* Researches and Transactions of the New York State Archeological Association. Rochester, N.Y.: New York State Archeological Association.

———. 1940. *Two Prehistoric Village Sites at Brewerton, N.Y.* Rochester Museum of Arts and Sciences Research Records 5. Rochester, N.Y.

———. 1957. *Traces of Early Man in the Northeast.* New York State Museum and Science Service Bulletin 358. Albany, N.Y.

———. 1959. *The Stony Brook Site and Its Relation to Archaic and Transitional Cultures on Long Island.* Rev. 2nd ed. New York State Museum and Science Service Bulletin 372. Albany, N.Y.

———. 1969. *The Archaeology of New York State.* Garden City, N.Y.: Natural History Press.

———. 1971. *New York Projectile Points: A Typology and Nomenclature.* New York State Museum Bulletin 384. Albany, N.Y.

Ritchie, William A., and Robert E. Funk. 1971. "Evidence for Early Archaic Occupations on Staten Island." *Pennsylvania Archaeologist* 41(3): 45–59.

———. 1973. *Aboriginal Settlement Patterns in the Northeast.* New York State Museum and Science Service Memoir 20. Albany, N.Y.

Ritchie, William A., and Richard S. MacNeish. 1949. "The Pre-Iroquoian Pottery of New York State. *American Antiquity* 15(2): 97–124.

Robbins, Maurice. 1980. *Wapanucket*. Attleboro, Mass.: Massachusetts Archaeological Society, Inc.

Robinson, Brian S. 1976–1977. "Seabrook Tidal Marsh Site: A Preliminary Report." *New Hampshire Archeologist* 19: 1–7.

———. 1983. "The Seabrook Station (Rocks Road) Site." *New Hampshire Archeological Society Newsletter* (June): 2–3.

———. 1985. "The Nelson Island and Seabrook Marsh Sites: Late Archaic, Marine Oriented People on the Central New England Coast." In *Ceramic Analysis in the Northeast*, ed. James B. Petersen. Occasional Publications in Northeastern Anthropology 9. Rindge, N.H.: Department of Anthropology, Franklin Pierce College. 1–107.

———. 1992. "Early and Middle Archaic Period Occupation in the Gulf of Maine Region: Mortuary and Technological Patterning." *Occasional Publications in Maine Archaeology* 9: 63–116.

Robinson, Brian S., and Charles E. Bolian. 1987. "A Preliminary Report on the Rocks Road Site (Seabrook Station): Late Archaic to Contact Period Occupation in Seabrook, New Hampshire." *New Hampshire Archeologist* 28(1): 19–51.

Robinson, Brian S., James B. Petersen, and Ann K. Robinson, eds. 1992. *Early Holocene Occupation in Northern New England: Occasional Publications in Maine Archaeology* 9. Augusta, Maine: Maine Historic Preservation Commission.

Rolando, Victor R. 1993. "The New Hampshire Iron Works, Franconia, New Hampshire, 1805–1864." *Society for Industrial Archeology—New England Chapters Newsletter* 13(2): 3–5.

———. 1996. *Report on the Surface Remains of the New Hampshire Iron Factory Company at Franconia, New Hampshire*. Privately printed by Victor R. Rolando.

Sands, John O. 1996. "Gunboats and Warships of the American Revolution." In *Ships and Shipwrecks of the Americas*, ed. George F. Bass. New York: Thames and Hudson. 149–168.

Sanger, David, William R. Belcher, and Douglas C. Kellogg. 1992. "Early Holocene Occupation at the Blackman Stream Site, Central Maine." *Occasional Publications in Maine Archaeology* 9: 149–161.

Sargent, Howard. 1950. "Summary of the First Season's Work at Clark's Island." *New Hampshire Archeological Society Newsletter* 2.

———. 1951. "Preliminary Report on the Excavations at Clark's Island." *New Hampshire Archeological Society Newsletter* (January).

———. 1959. "The Pickpocket Falls Site." *New Hampshire Archeologist* 9: 2–6.

———. 1969. "Prehistory in the Upper Connecticut Valley." In *An Introduction to the Archaeology and History of the Connecticut Valley Indian*, ed. William R. Young. Springfield, Mass.: Springfield Museum of Science. 1(1): 28–32.

———. 1971. "Prehistory in the Upper Connecticut Valley." *New Hampshire Archeologist* 16: unpaginated.

———. 1975. "Preservation History: The Archeological Record." *New Hampshire Archeologist* 18: 18–23.

———. 1980. *Archeological Salvage at the Weirs, Laconia, New Hampshire*. Concord: New Hampshire Water Supply and Pollution Control Commission.

———. 1982. *A New Look at an Old Lake: Monadnock Perspectives*. West Peterborough, N.H.: Monadnock Perspectives, Inc.

———. 2003–2004. "A New Look at an Old Lake." *New Hampshire Archeologist* 43/44(1): 12–18.

Sargent, Howard, and Francois G. Ledoux. 1973. "Two Fluted Points from New England." *Man in the Northeast* 5: 67–68.

Schoolcraft, Henry R. 1851–1857. *Historical and Statistical Information Respecting the History, Condition, and Prospects of the Indian Tribes of the United States*. Philadelphia: Lippincott, Grambo. 6 vols.

Skinas, David C. 1981. "The Wadleigh Falls Site (NH39-2): A Preliminary Report of the 1980 Excavations." *New Hampshire Archeologist* 22(1): 16–30.

Snow, Dean R. 1977. "The Archaic of the Lake George Region." In *Amerinds and Their Paleoenvironments in Northeastern North America*, ed. Walter Newman and Bert Salwen. Annals of the New York Academy of Sciences 288: 431–438. New York: New York Academy of Sciences.

———. 1980. *The Archaeology of New England*. New York: Academic Press.

———. 1993. "Obituary of Howard R. Sargent, 1922–1993." *Man in the Northeast* 45: 1–2.

South, Stanley, ed. 1994. *Pioneers in Historical Archaeology: Breaking New Ground*. New York: Plenum Press.

Spiess, Arthur E., ed. 1978. *Conservation Archaeology in the Northeast: Toward a Research Orientation*. Cambridge, Mass.: Harvard University. Peabody Museum Bulletin 3.

Spiess, Arthur E., Mary Lou Curran, and John R. Grimes. 1984–1985. "Caribou (*Rangifer tarandus L.*) Bones from New England Paleo-Indian Sites." *North American Archaeologist* 6(2): 145–159.

Spiess, Arthur E., and Deborah Wilson. 1987. *Michaud: A Paleoindian Site in the New England–Maritimes Region*. Occasional Publications in Maine Archaeology 6. Augusta, Maine: Maine Historic Preservation Commission and the Maine Archaeological Society.

Starbuck, David R. 1976–1977. "A Progress Report on the New England Glassworks Project." *New Hampshire Archeologist* 19: 25–34.

———. 1977. "An Archeological Assessment of the Proposed Bicentennial Square in Concord, New Hampshire, Phase II." Manuscript prepared for the State Historic Preservation Office, Concord, N.H.

———. 1978. "The Excavation of the New England Glassworks in Temple, New Hampshire." In *New England Historical Archeology*, ed. Peter Benes. Dublin Seminar for New England Folklife: Annual Proceedings, 1977. Boston: Boston University Scholarly Publications. 75–85.

———. 1980a. "The Archeology of Canterbury Shaker Village." *New Hampshire Archeologist* 21: 67–79.

———. 1980b. "The Middle Archaic in Central Connecticut: The Excavation of the Lewis-Walpole Site (6-HT-15). In *Early and Middle Archaic Cultures in the Northeast*, ed. David R. Starbuck and Charles E. Bolian. *Occasional Publications in Northeastern Anthropology* 7. Rindge, N.H.: Department of Anthropology, Franklin Pierce College. 5–37.

———. 1982a. "Excavations at Sewall's Falls (NH31-30) in Concord, N.H." *New Hampshire Archeologist* 23: 1–36.

———. 1982b. *A Middle Archaic Site: Belmont, New Hampshire.* Concord: New Hampshire Department of Public Works and Highways.

———. 1983a. "The New England Glassworks in Temple, New Hampshire." *IA: The Journal of the Society for Industrial Archeology* 9(1): 45–64.

———. 1983b. "Survey and Excavation along the Upper Merrimack River in New Hampshire." *Man in the Northeast* 25: 25–41.

———. 1984a. "New Hampshire's Earliest Glass Factory: The New England Glassworks, 1780–1782." *Historical New Hampshire* 39(1–2): 45–63.

———. 1984b. "Further Excavations at Sewall's Falls (NH31-30)." *New Hampshire Archeologist* 25(1): 1–11.

———. 1984c. "The Shaker Concept of Household." *Man in the Northeast* 28: 73–86.

———. 1985a. "The Garvins Falls Site (NH37-1): The 1982 Excavations." *New Hampshire Archeologist* 26(1): 19–42.

———. 1985b. "The Industrial Archeology of New Hampshire." *Historical New Hampshire* 40(1–2): 84–99.

———. 1985c. "Three Seasons of Site Survey and Excavation at Sewall's Falls (NH31-30)." *New Hampshire Archeologist* 26(1): 87–102.

———. 1985d. "West Bank at Sewall's Falls." *New Hampshire Archeological Society Newsletter* 1(2): 9, 11.

———. 1986a. "The New England Glassworks: New Hampshire's Boldest Experiment in Early Glassmaking." Special issue. *New Hampshire Archeologist* 27(1): 1–148.

———. 1986b. "The Shaker Mills in Canterbury, New Hampshire." *IA: The Journal of the Society for Industrial Archeology* 12(1): 11–38.

———. 1988a. "Documenting the Canterbury Shakers." *Historical New Hampshire* 43(1): 1–20.

———. 1988b. "America's Oldest Summer Place." *Archaeology* 41(6): 60–61.

———. 1988c. "John Wentworth's Frontier Plantation in Wolfeboro, New Hampshire." *Historical New Hampshire* 43(3): 181–201.

———. 1989. "America's First Summer Resort: John Wentworth's 18th-Century Plantation in Wolfeboro, New Hampshire." *New Hampshire Archeologist* 30(1).

———. 1990a. "Canterbury Shaker Village: Archeology and Landscape." *New Hampshire Archeologist* 31(1): 1–163.

———. 1990b. "Those Ingenious Shakers!" *Archaeology* 43(4): 40–47.

———. 1990c. "The Timber Crib Dam at Sewall's Falls." *IA: The Journal of the Society for Industrial Archeology* 16(2): 40–61.

———. 1991. "The Lewis-Walpole Site (6-HT-15)." *New Hampshire Archeologist* 32(1): 73–86.

———. 1994a. "An Introduction to New Hampshire Industrial Archeology." *IA: The Journal of the Society for Industrial Archeology* 20(1–2): 4–18.

———. 1994b. "The Cog Railway on Mount Washington." *IA: The Journal of the Society for Industrial Archeology* 20(1–2): 101–118.

———. 1997. "Recent Excavations at Canterbury Shaker Village." *New Hampshire Archeologist* 37(1): 9–27.

———. 1998. "New Perspectives on Shaker Life." *Expedition* 40(3): 3–16.

———. 1999a. *The Great Warpath*. Hanover, N.H.: University Press of New England.

———. 1999b. "Latter-Day Shakers." *Archaeology* 52(1): 28–29.

———. 2000a. "A Trash Pit Behind the Squire Abiathar Britton House in Orford." *New Hampshire Archeologist* 40(1): 43–62.

———. 2000b. "Waiting for the Second Coming: The Canterbury Shakers; An Archaeological Perspective on Blacksmithing and Pipe Smoking." *Northeast Historical Archaeology* 29: 83–106.

———. 2002. *Massacre at Fort William Henry*. Hanover, N.H.: University Press of New England.

———. 2003–2004. "New Hampshire's Best-Loved Industrial Site: The Cog Railway on Mount Washington." *New Hampshire Archeologist* 43/44(1): 107–118.

———. 2004. *Neither Plain nor Simple*. Hanover, N.H.: University Press of New England.

Starbuck, David R., and Charles E. Bolian, eds. 1980. *Early and Middle Archaic Cultures in the Northeast*. Occasional Publications in Northeastern Anthropology 7. Rindge, N.H.: Department of Anthropology, Franklin Pierce College.

Starbuck, David R., and Mary Dupre. 1985a. "The Hazeltine Pottery Site, Concord, N.H. (NH37-8)." *New Hampshire Archeologist* 26(1): 135–145.

———. 1985b. "Production Continuity and Obsolescence of Traditional Red Earthenwares in Concord, New Hampshire." In *Domestic Pottery of the Northeastern United States: Regional Production and World Trade*, ed. Sarah Peabody Turnbaugh. Orlando: Academic Press. 133–152.

Stewart-Smith, David. 1994. "The Pennacook: Lands and Relations; An Ethnography." *New Hampshire Archeologist* 33/34(1): 66–80.

Stinson, Wesley R. 1991. "Souhegan River Survey." *New Hampshire Archeological Society Newsletter* 7(1): 7–8.

Stone, Robert E. 1971. "Megalithic Mystery Hill." *Northeast Historical Archaeology* 1(1): 22–23.

Switzer, David C. 1976. "The Penobscot Bay Project: Revolutionary War Wrecks." *Archaeology* 29(3): 204–209.

———. 1981. "Nautical Archaeology in the Penobscot Bay: The Revolutionary Privateer *Defence*." In *Aspects of Naval History*, ed. Craig Symonds. Annapolis: U.S. Naval Institution Press. 90–101.

———. 1983. "The Excavation of the Privateer *Defence*." *Northeast Historical Archaeology* 12: 43–50.

———. 1985. "Archeology under New Hampshire Waters: The Present and the Future." *Historical New Hampshire* 40(1–2): 34–56.

———. 1991. "The Hart's Cove Wreck: Nautical Archaeology in New Hampshire Waters." *Seaways: Journal of Maritime History and Research* 2(3): 27–32.

———. 1994. "Nautical Archeology in Hart's Cove." *New Hampshire Archeologist* 33/34: 97–104.

———. 2003–2004. "The Schooner *Lizzie Carr*: A Legacy of the Last Days of American Sail." *New Hampshire Archeologist* 43/44(1): 93–106.

Switzer, David C., and Brendan P. Foley. 2001. "Underwater Study of Enfield Shaker Bridge." New Hampshire Department of Transportation. Project no. BRO-X-145(003), 12967.

Taylor, William B. 1976. "A Bifurcated Point Concentration." *Bulletin of the Massachusetts Archaeological Society* 37(3–4): 36–41.

Taylor, William L. 1984. "The Concord (New Hampshire) Gasholder: Last Intact Survivor from the Gas-Making Era." *IA: The Journal of the Society for Industrial Archeology* 10(1): 1–16.

Thomas, David Hurst. 1994. *Exploring Ancient Native America*. New York: Macmillan.

———. 2002. *Skull Wars*. New York: Basic Books.

Thomas, Peter A. 1973a. "Jesuit Rings: Evidence of French-Indian Contact in the Connecticut River Valley." *Historical Archaeology* 7: 54–57.

———. 1973b. "Squakheag Ethnohistory: A Preliminary Study of Culture Conflict on the Seventeenth-Century Frontier." *Man in the Northeast* 5: 27–36.

———. 1979. "In the Maelstrom of Change: The Indian Trade and Cultural Process in the Middle Connecticut River Valley: 1635–1665." Ph.D. dissertation, University of Massachusetts, Amherst.

———. 1992. "The Early and Middle Archaic Periods as Represented in Western Vermont." *Occasional Publications in Maine Archaeology* 9: 187–203.

Thomas, Peter A., and Brian S. Robinson. 1983. *The John's Bridge Site: VT-FR-69; An Early Archaic Period Site in Northwestern Vermont*. Burlington: Vermont Archaeological Society.

Thompson, Edward Herbert. 1932. *People of the Serpent: Life and Adventure among the Mayas*. Boston: Houghton Mifflin Company.

Trigger, Bruce G., ed. 1978. *Handbook of North American Indians*. Vol. 15, *Northeast*. Washington, D.C.: Smithsonian Institution.

Vescelius, Gary S. 1955. "The Antiquity of Pattee's Caves." Reprinted in *New England Antiquities Research Association Journal* 17(1): 2–16, 17(2): 28–42, 17(3): 59–67.

———. 1956. "Excavations at Pattee's Caves." *Bulletin of the Eastern States Archaeological Federation* 15: 13–14.

Vitelli, Karen D. 1996. *Archaeological Ethics*. Lanham, Md.: AltaMira Press.

Waldbauer, Richard C. 1985. "Material Culture and Agricultural History in New Hampshire." *Historical New Hampshire* 40(1–2): 61–71.

———. 1986. "House Not a Home: Hill Farm Clustered Communities." *Man in the Northeast* 31: 139–150.

Watkins, Lura Woodside. 1950. *Early New England Potters and Their Wares*. Cambridge, Mass.: Harvard University Press.

Webber, Laurence. 1973. "New Hampshire's First Ironworks." *New Hampshire Profiles* (June): 44–49.

Welch, Sarah N. 1972. *A History of Franconia, New Hampshire*. Littleton, N.H.: Courier Printing Company, Inc.

Wheeler, Kathleen. 1992. "The Characterization and Measurement of Archaeological Depositional Units: Patterns from Nineteenth-Century Urban Sites in Portsmouth, New Hampshire." Ph.D. dissertation, University of Arizona.

———. 1999. "Contributions of Women to the Acquisition, Maintenance, and Discard of Portable Estates." *Northeast Historical Archaeology* 28: 41–56.

———. 2000. "Theoretical and Methodological Considerations for Excavating Privies." *Historical Archaeology* 34(1): 3–19.

White, Carolyn L. 2002. "Constructing Identities: Personal Adornment from Portsmouth, New Hampshire, 1680–1820." Ph.D. dissertation, Boston University.

———. 2004. "What the Warners Wore: An Archaeological Investigation of Visual Appearance." *Northeast Historical Archaeology* 33: 39–66.

———. 2005. *American Items of Personal Adornment, 1680–1820: A Guide to Identification and Interpretation.* Lanham, Md.: AltaMira Press.

Wilding-White, Sherry. 1994. "The Abbot-Downing Company and the Concord Coach." *IA: The Journal of the Society for Industrial Archeology* 20(1–2): 89–90.

Wilson, John S. 1990. "We've Got Thousands of These! What Makes an Historic Farmstead Significant?" *Historical Archaeology* 24(2): 23–33.

Wingerson, Roberta. 1994. "The Mill Village on Goose Creek: Harrisville, New Hampshire." *IA: The Journal of the Society for Industrial Archeology* 20(1–2): 91–100.

Winter, Eugene. 1975. "The Smyth Site at Amoskeag Falls: A Preliminary Report." *New Hampshire Archeologist* 18: 5–8.

———. 1985. "The Garvin's Falls Site (NH37-1): The 1963–1970 Excavations." *New Hampshire Archeologist* 26(1): 1–18.

Witthoft, John. 1952. "A Paleo-Indian Site in Eastern Pennsylvania: An Early Hunting Culture." *Proceedings of the American Philosophical Society* 96(4): 464–495.

Index